D0991900

ELECTRICAL INSTALLATION TECHNOLOGY AND PRACTICE

GENERAL TECHNICAL SERIES

General Editor:
AIR COMMODORE J. R. MORGAN, O.B.E.
B.Sc. (Eng.), M.I.Mech.E., F.R.Ae.S., R.A.F. (Ret'd)
Formerly
Director of Studies
Royal Air Force Technical College
and
Deputy Director, Educational Services, Air Ministry

RADIO
J. D. Tucker, A.M.Brit.I.R.E. and
D. F. Wilkinson, B.Sc. (Eng.), A.M.I.E.E.
(*In three volumes*)

MATHEMATICS FOR TELECOMMUNICATIONS AND ELECTRICAL ENGINEERING
W. H. Grinsted, O.B.E., M.I.E.E., F.C.G.I. and
D. G. Spooner, B.Sc. (Lond.)
(*In two volumes*)

ENGINEERING SCIENCE AND CALCULATIONS
W. E. Fisher, O.B.E., D.Sc., A.M.I.Mech.E.
(*In three volumes*)

BUILDING CRAFT SCIENCE
(*Intermediate*)
A. W. Nobbs, B.Sc.

ELEMENTARY TELECOMMUNICATION PRACTICE
J. R. G. Smith, A.M.I.E.E.

CONCISE PRACTICAL SURVEYING
W. G. Curtin, A.M.I.Struct.E. and R. F. Lane, F.R.I.C.S.

SKETCHING FOR CRAFTSMEN
E. Hoyle, M.B.E.

SIMPLIFIED CALCULUS
Instructor Captain F. L. Westwater, O.B.E., R.N. (Ret'd),
M.A. (Edin.), M.A. (Cantab), A.M.I.E.E.

ENGINEERING DRAWING AND MATERIALS
For Mechanical Engineering Technicians
H. Ord, A.M.I.E.D. A.R.Ae.S., F.R.Econ.S.

ELECTRICAL INSTALLATION TECHNOLOGY AND PRACTICE

J. O. PADDOCK
M.A.S.E.E., H.N.C., C.G.L.I. Full Tech. Cert.
Department of Engineering and Science,
South East Kent Technical College.

R. A. W. GALVIN
A.M.A.S.E.E., Grad. I.E.E., C.G.L.I. Cert. in Elect. Eng. Pract.
Department of Engineering and Science,
South East Kent Technical College.

THE ENGLISH UNIVERSITIES PRESS LTD.
102 NEWGATE STREET
LONDON · E.C.1

First printed 1964

Printed and bound for
The English Universities Press Ltd.
by Cox and Wyman Ltd.,
London, Reading and Fakenham

Editor's Introduction

The new awareness of the imperative need to make the very most of our technical potential makes a foreword to this General Technical Series almost unnecessary, for it aims directly at encouraging young men – and women – to extend their interest, widen their knowledge, and improve their technical skills.

The City and Guilds of London Institute makes special provision for the technician to acquire a qualification appropriate to his Craft. The wide range of examinations now held under its auspices is ample evidence not merely of the need to cater for the technician but also of the growing desire of the Craftsman to improve his knowledge of his Craft. Many of the books in the present series will be related to syllabuses of the City and Guilds of London Institute, but this will not limit their use merely to preparation for the examinations held by that body. The aim is to encourage students to study those technical subjects which are closely related to their daily work and, by so doing, to obtain a better understanding of basic principles. Any study of this kind cannot fail to stimulate interest in the subject and should produce a technician with a clearer understanding of what he is doing and how it should best be done.

But although the series is intended to appeal, in the first instance, to students who are interested in the certificates offered by the City and Guilds of London Institute, that must be regarded as only the immediate aim. Those students who, as a result of their initial endeavours, find that they are capable of going further should aim at obtaining either a National Certificate in an appropriate field of engineering or, alternatively, a General Certificate of Education at a level appropriate to their potential attainment.

All the books in the series will be written by experienced and well qualified teachers who are thoroughly conversant with the problems encountered by young men and women in studying the subjects with which their books deal.

J. R. M.

Preface

This book is primarily intended for students preparing to take the City and Guilds of London Institute's Course B Examination in Electrical Installation Work, but it will also be useful to those following Electrical Installation Craft Practice, Electrical Fitters and Electrical Technicians Courses. Each chapter of the book is followed by a selection of workshop exercises designed to give practical insight into the topics covered in the chapter. It is most important for students to be able to prepare accurate and concise reports of the practical work which they carry out, as their practical knowledge is assessed by means of written examinations. To this end students are advised to maintain a 'log-book' in which they record every practical exercise which they perform. Constructional details of the demonstration boards needed for certain exercises are included in Chapter 14, as this type of apparatus is not generally available from commercial sources.

The authors are indebted to Mr. A. F. J. Riley for his invaluable assistance in preparing the illustrations for this book, and to their colleague, Mr. W. A. Baker, A.M.A.S.E.E., for his help in reading the proofs.

The questions from recent examination papers given in Chapter 15 have been reproduced by kind permission of the following examining bodies:

City and Guilds of London Institute (C.G.L.I.).
East Midland Educational Union (E.M.E.U.).
Northern Counties Technical Examinations Council (N.C.T.E.C.).
Union of Lancashire and Cheshire Institutes (U.L.C.I.).
Welsh Joint Education Committee (W.J.E.C.).

Acknowledgement is also made to the Institution of Electrical Engineers for permission to quote from the Regulations for the 'Electrical Equipment of Buildings' and to Messrs. Enfield Standard Power Cables Limited for permission to reproduce drawings of their catenary supported wiring system.

Finally the authors wish to thank the staff of The English Universities Press Limited for their encouragement and many helpful suggestions in the preparation of this book.

Dover, 1964

J. O. PADDOCK
R. A. W. GALVIN

Contents

Introduction

What the Electrical Installation Provides

1. The electricity supply authorities are responsible for providing a supply of electricity to suitable terminals on a consumer's premises; the electrical installation in the premises provides the means of conveying the electricity to the equipment where it is to be used. Like fire, electricity is a good servant but a bad master and so, before anyone can install a safe and efficient electrical system, it is essential for him to be familiar with the nature of electricity and the dangers inherent in its use.

2. The two main hazards involved, wherever electricity is employed, are the danger of shock and the danger of fire. Both types of risk may be reduced to negligible proportions by using suitable materials and correct methods of installation. Because of the vital need to maintain high standards in carrying out installation work various lists of regulations, requirements and codes of practice are published, some are enforceable by law while others, although being merely recommendations are, nevertheless, generally accepted as setting the standards to which every installation should be constructed.

3. The more important sets of regulations concerning electrical installation work are listed below; no electrician can claim to be fully competent if he is not familiar with the contents of these publications.

Electricity Supply Regulations

4. These regulations, issued by the Minister of Fuel and Power, give the supply undertakings mandatory powers to insist on certain minimum standards of installation work before they need provide a supply to a consumer. In general if an installation meets the requirements of the I.E.E. Wiring Regulations then it will also satisfy the requirements of the Electricity Supply Regulations.

Electricity (Factories Act) Special Regulations, 1908 and 1944

5. All electrical equipment installed in factories and workshops must comply with the requirements of the Factories Act. Many of the requirements are similar to those of the I.E.E. wiring regulations but there are certain additional requirements, particularly for conditions where special hazards exist. Anyone engaged in factory installation work should obtain and study the 'Memorandum by the

Senior Electrical Inspector of Factories on the Electricity Regulations' (H.M.S.O.) which gives a lucid explanation of this rather complex subject.

Regulations for the Electrical Equipment of Buildings

6. These regulations, issued by the Institution of Electrical Engineers, provide a comprehensive list of the requirements which experience has shown to be necessary for a safe and efficient installation. Although they are not in themselves mandatory an installation which complies with these regulations will normally also meet the requirements of the various other mandatory regulations. Since the I.E.E. Regulations are so comprehensive they can be said to constitute the 'electricians' bible' and every one concerned with electrical contracting should be thoroughly familiar with them. In view of the importance of these regulations, summaries of their requirements have been included whenever appropriate in this book, nevertheless students are strongly recommended to obtain their own copies of the regulations so that they can compare the full text of any particular regulation with the summaries given.

CHAPTER 1

Electric Cables and Joints

CONDUCTORS AND INSULATORS

1.1 Conductors

Any material which will allow the free passage of an electric current is known as a conductor. Conducting materials vary in the degree to which they can conduct electricity; good conductors are required for connecting leads in electric circuits so that they may convey the current with a minimum loss of voltage. Materials giving a somewhat higher resistance are sometimes needed for controlling currents, e.g. for the construction of rheostats, motor starters, etc.; similar materials are also required for the construction of heating elements where heat is produced by forcing a current through a relatively high resistance. Some typical materials used as conductors in electric circuits are listed below.

(a) *Silver*. This is the best-known conductor but it is too expensive for general use. The contacts of some switches are plated with silver to lower the contact resistance.

(b) *Copper*. This material is widely used for the manufacture of electric wires, cables and bus bars. Its conductivity is second only to silver and it is reasonably priced. As an electrical conductor it has the following advantages:

(i) Low resistance.
(ii) It is ductile and therefore easily formed into wires.
(iii) It is readily 'tinned' for soldering.

(c) *Aluminium*. This is coming into increasing use as an electrical conductor since, although it is not such a good conductor as copper, its light weight is an advantage in many situations.

(d) *Brass*. Brass is often used for the manufacture of terminals and various parts of electric fittings. Its advantages are:

(i) It is harder than copper.
(ii) It is easily machined.

(iii) It can be readily cast.
(iv) Like copper it is also easily 'tinned' for soldering.

(e) *Nichrome*. This type of resistance wire is used for many purposes of which the following are typical:

(i) Manufacture of fixed and variable resistors.
(ii) Heating elements.

(f) *Eureka* (Constantan or Advance). This is a better quality resistance wire, as its resistance value does not change appreciably with temperature. It is widely used in the construction of electrical instruments.

(g) *Manganin*. Manganin is a high quality resistance wire, which is even more constant in resistance value than Eureka. It is expensive and requires special heat treatment to develop its good properties and it is used mainly for precision resistors as used in laboratories.

(h) *Tungsten*. This material has a very high melting point and so is used in the manufacture of electric lamp filaments.

(i) *Carbon*. Although this is not a metal it is a fairly good conductor and its physical properties make it very suitable for use as 'brushes' in electrical machines. Carbon can be mixed with clay and other materials for the manufacture of carbon composition resistors; these can have very high values of resistance despite their small size and are widely used in radio and electronic equipment.

1.2 Insulators

Any material which does not allow the free passage of an electric current is known as an insulator. Insulators are used to confine electric currents to the conductors in which they are intended to flow, and to prevent leakage of electricity to adjacent conducting materials which are not intended to become 'alive'. Insulation is also needed to prevent 'short circuits' between various parts of an installation. Some typical insulating materials are listed below.

(a) *Rubber*. Widely used for covering wires and cables.
Some advantages:

(i) Good insulator.
(ii) Impervious to water.
(iii) Flexible.

Some disadvantages:

(i) Adversely affected by sunlight.
(ii) Not fireproof.

(b) *Poly-Vinyl-Chloride (P.V.C.)*. A modern thermoplastic material which is often used as an alternative to rubber. Some advantages and disadvantages as compared with rubber are:

Advantages:

(i) Resists chemical action and direct sunlight.
(ii) Not so inflammable.

Disadvantages:

(i) Does not afford the same degree of mechanical protection.
(ii) More expensive.

(c) *Paper*. Impregnated paper is often used to insulate the conductors in underground cables; it must be protected from the ingress of moisture.

(d) *Mineral Insulation*. Magnesium oxide is used as insulation in certain types of cables. It is extremely heat resistant but must be protected against ingress of moisture.

(e) *Mica*. Used for insulation where high temperatures are involved, e.g. heating elements. It is also widely used for insulation between the copper segments of the commutators of electrical machines. Its main disadvantage is its brittleness.

(f) *Asbestos*. Used to insulate connecting leads where high temperatures are involved.

(g) *Paxolin*. Used for insulating panels and barriers in switchgear, and other places where a rigid insulating sheet is required.

(h) *Bakelite*. Used for moulded insulating parts of electrical fittings.

CABLES

1.3 The Three Main Parts of a Cable

Cables used in electrical circuits are of many types but all consist of the following main parts:

(*a*) Conductor.
(*b*) Insulation.
(*c*) Mechanical protection.

(a) *Conductors* are usually made of copper, the conducting cores being formed from strands of copper wire so that the cable is more flexible than if solid cores were used. If vulcanized rubber insulation is to be used the copper conductors are tinned to prevent corrosion of the copper by the sulphur which is present in vulcanized rubber.

The size of cables used in domestic installations is normally stated as: No. of strands/Diameter of each strand; e.g. a cable commonly used for lighting circuits is 3/·029; this means that the cable consists of three strands each of ·029 in. diameter. For larger types of cable, the effective cross-sectional area of the core is often quoted as the size.

(b) *Insulation* of cables used in domestic installations is normally of vulcanized rubber (V.R.I.) or poly-vinyl-chloride (P.V.C.). Where mineral insulation (magnesia) is employed the cable has a copper outer sheath, this type of cable being known as Mineral Insulated Copper-sheathed Cable (M.I.C.S.).

(c) *Mechanical Protection* is provided to prevent damage to the cable during installation and throughout its subsequent service. The following types of cables, which are in common use, are discussed in more detail in Chapter 3.

(i) Vulcanized India-rubber Insulated (V.R.I.).
(ii) Flexible Cables and Cords.
(iii) Tough Rubber-sheathed (T.R.S.).
(iv) Lead-sheathed V.R.I.
(v) Poly-Vinyl-Chloride Insulated and Sheathed (P.V.C.).
(vi) Mineral Insulated (M.I.C.S.).
(vii) Paper Insulated Lead-covered (P.I.L.C.).

1.4. Stripping of Cables

The mechanical protection and insulation must be removed from the end of the cable to leave a suitable length of exposed conductor whenever it is required to connect the cable to a terminal or to make a joint.

The following points should be carefully observed when stripping cables:

(*a*) The strands forming the core must not be cut or nicked in any way as this could result in a loss of conducting area, reduction of current carrying capacity and increased resistance. A 'nicked' conductor is always a potential source of trouble as it may easily break completely at a later date.

(*b*) If any cable has an outer protection of cloth, tape or braiding, such as V.R.I. which usually has both, this must be removed to expose at least half an inch of insulation. This is because cloth is not a good insulator and can be a cause of a low insulation resistance.

(*c*) The insulation of the core must not be damaged when removing the outer covering. This particularly applies to multi-core cables such as P.V.C. and T.R.S.

JOINTS

1.5 Mechanical Joints

(a) *Conduit and trunking systems.* When an installation is wired using single core cables enclosed by steel conduit or trunking, it is good practice to avoid joints by using the 'loop-in' system of wiring as explained in Chapter 2. If, however, a joint cannot be avoided this must be made by means of either a mechanical connector or by a soldered joint. The joint must be made in a readily accessible joint box. Fig. 1.1 shows such a joint made using a porcelain shrouded connector, accommodated in a conduit box.

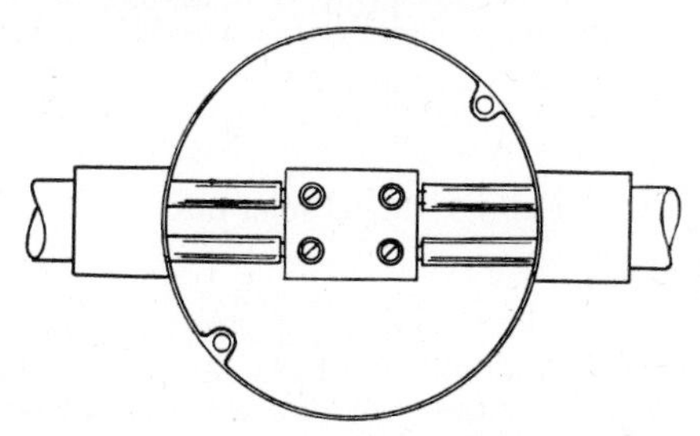

FIG. 1.1 Porcelain Connector used in Conduit Box

(b) *All-insulated systems.* When an installation is wired using T.R.S. or P.V.C. cable, joints are effected using all-insulated joint boxes, which usually possess four fixed terminals as shown in Fig. 1.2.

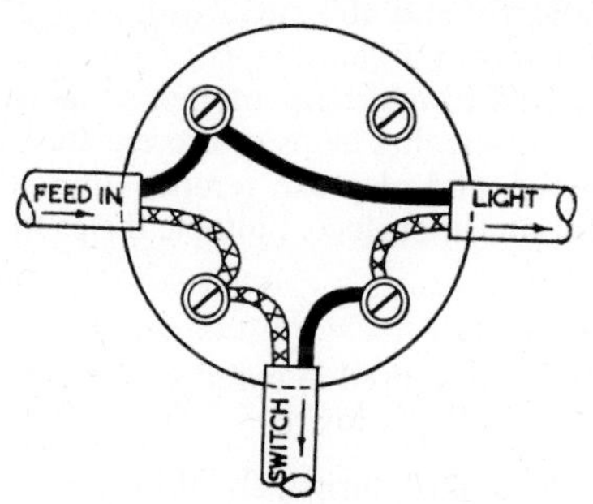

FIG. 1.2 All-insulated Joint Box

(c) *Lead-sheathed wiring systems.* The joint boxes used in connection with lead-sheathed wiring systems are made of metal and are fitted with arrangements for bonding the lead sheaths in order to maintain the earth continuity of the installation. The joints between the conductors are made either using porcelain shrouded connectors or scruits as shown in Fig. 1.3.

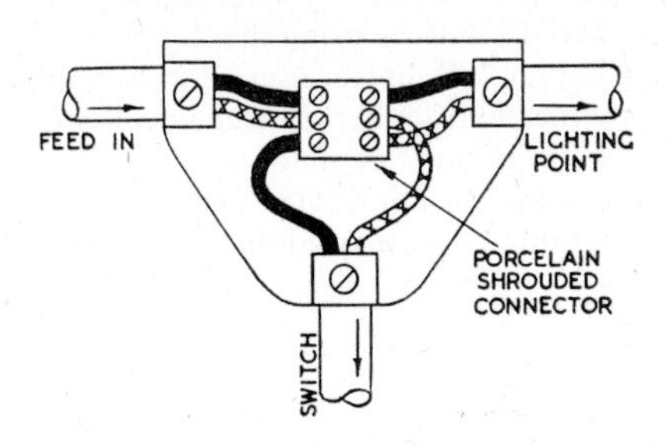

FIG. 1.3 Metal Joint Box

1.6 Soldered Joints

(a) A correctly made soldered joint possesses a low resistance which gives good electrical continuity and has a fair degree of mechanical strength. In a soundly made joint the two parts to be joined are linked by a thin film of solder which has penetrated into the surface of each metal.

Solder is an alloy of tin and lead which melts at a comparatively low temperature. For most electrical joints the grade of solder known as 'Tinmans' solder is most suitable. This is an alloy containing approximately 60% tin and 40% lead with a small amount of antimony added; it melts at about 200° C.

(b) Flux is required to assist the molten solder to flow over the surface of the metal. The flux helps to prevent the surface absorbing oxygen from the air, and also can remove any oxides which may already be present on the surface. Suitable fluxes for electrical work are:

(i) Pure amber resin.
(ii) 'Activated' resin as used in cored solders.
(iii) Solder paste such as Fluxite.

Acid fluxes, such as 'killed spirits', should not be used as they tend to cause corrosion if the part concerned is not thoroughly washed after soldering, and this is seldom practical with electrical work. The amount of flux used should be just sufficient to encourage the solder to flow readily; excessive fluxing only results in a messy job, without improving the soldered joint. If a large quantity of flux is needed before the solder will run and tin properly, this points to the fact that the parts have not been sufficiently cleaned before starting to solder them.

(c) The secret of successful soldering lies in the cleaning of the parts to be soldered. All traces of oil or grease must be removed and the metal thoroughly cleaned using emery paper if necessary. Copper and brass are easy metals to solder, iron is more difficult and, correspondingly, more care must be taken in preparation. Wherever possible the parts should be 'tinned', that is covered with a thin layer of solder before the main soldering job is begun. Aluminium can only be soldered by using special techniques which it is beyond the scope of this book to discuss.

(d) The part to be soldered must be heated until solder will flow freely over its surface. Insufficient heating may result in a layer of solder over the metal which is not actually alloyed into the surface of the metal. This solder can be easily peeled off and it *does not* make a good electrical connection. Joints with this type of defect are called 'dry joints'. On the other hand excessive heating can cause damage to the insulation of a cable due to heat conducted along the wires. Methods of soldering vary mainly in the way in which heat is applied to the job. The principal methods used by electricians are:

(i) Soldering iron; this tool has a copper bit which may be heated by blowlamp, gas, electrically or by any other convenient method.
(ii) Metal pot; an iron pot or ladle containing solder is heated and the molten solder is subsequently poured over the job.
(iii) Blowlamp; the part to be soldered may be directly heated using a blowlamp.

Examples of the use of these methods will be found in exercises Nos. 1.9 and 1.10. While one method only has been described for each job it must be emphasized that any of the jobs could be carried out using any of the methods. The particular method used in any circumstance depends on the nature of the job; for example joints between wires are best made using a soldering iron or metal pot, since direct application of a blowlamp may cause excessive burning of the adjacent insulation; whereas when soldering a lug, direct heating by blowlamp or the use of a metal pot is advisable, as very often the size of the lug involved means that more heat is required than can be obtained from a normal sized soldering iron.

EXERCISES

The exercises which conclude this chapter are designed to give practice in the fundamental operations of preparing, terminating and jointing electric cables.

Exercise No. 1.1 — To prepare V.R.I. taped and braided cable for termination, and to join two lengths using a porcelain shrouded connector.

Materials

2 Lengths 3/·029 taped and braided cable.
Porcelain connector.

Diagrams

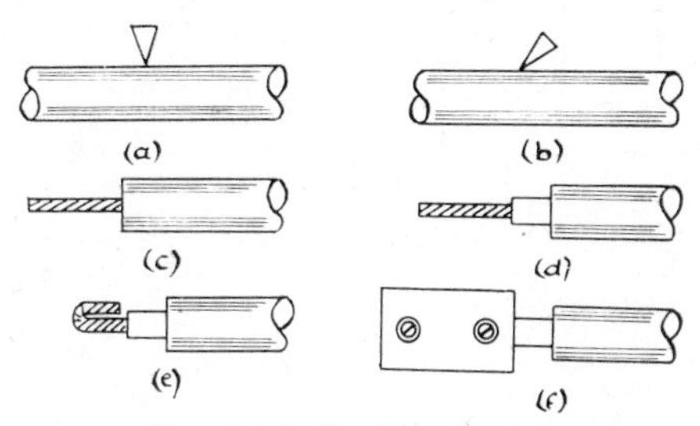

FIG. 1.4 (a–f) Exercise 1.1

(a) Knife blade held at incorrect angle; (b) Knife blade held at correct angle; (c) Tape braid and rubber removed; (d) Tape and braid removed a further ½ in.; (e) End tightly twisted and bent back; (f) Insulation close to connector.

Procedure

Prepare each cable as follows:

1. Remove the tape, braid and rubber from one end of the cable to leave a suitable length of exposed conductor (say ½ in.), using a sharp knife as shown in Fig. 1.4 (a–c).
2. Remove the tape and braid a further ½ in. from the end so that the insulation is exposed. (Fig. 1.4 (d).)
3. Twist the conductors tightly together in the direction of the 'lay' of the cable, and if necessary to fill the hole in the connector bend the end of the conductor back on itself. (Fig. 1.4 (e).)
4. Insert both conductors into the porcelain shrouded connector and tighten the grub screws. (Fig. 1.4 (f).)

Answer the following questions

1. What are the three main parts of a cable?
2. What is the reason for removing the tape and braid a further ½ in. from the end?
3. What could be the result of stripping the cable with the knife held at an incorrect angle?

EXERCISE No. 1.2

To connect a lamp holder to a ceiling rose using a flexible cord.

Materials

1 B.C. Lamp holder.
1 5A Two-plate ceiling rose.
14/·0076 Flexible cord.

Diagrams

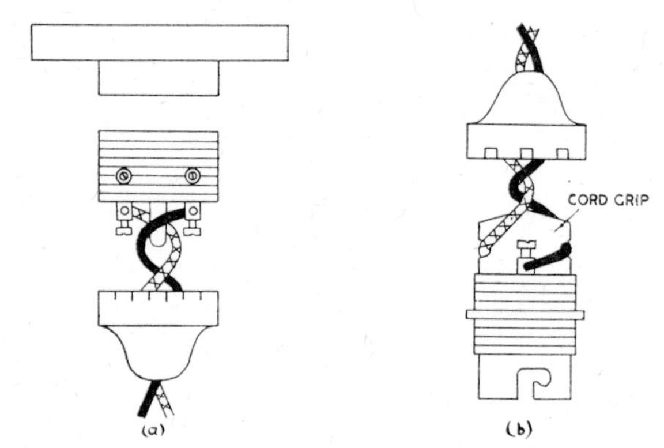

FIG. 1.5 (a–b) Exercise 1.2
(a) Ceiling rose; (b) Lamp holder

Procedure

1. Remove the braiding and insulation for a suitable length from both ends of the flexible cord.
2. Separate enough of the flexible cord at both ends to make connections to lamp holder and ceiling rose.
3. Connect one end of the flexible cord to the lamp holder as shown in the sketch, taking particular note of the following points:
   (*a*) Ensure that there are no loose strands of wire which are not inserted in the terminal.
   (*b*) Ensure that the flexible cord is securely held by the cord gripping arrangements.
4. Connect the other end of the flexible cord to the ceiling rose as shown in the sketch, once again ensure that there are no loose strands of wire and that the flexible cord is correctly secured by the cord gripping arrangements.

Answer the following questions

1. Why is it important that all strands of the flexible cord be securely held by the terminal?
2. Why are the cord gripping arrangements necessary?
3. Define (*a*) flexible cable, (*b*) flexible cord.

EXERCISE No. 1.3 — To connect a flexible cord to a 13A fused plug top.

Materials

70/·0076 Three-core flexible cord.
1 13A Fused plug top.

Diagram

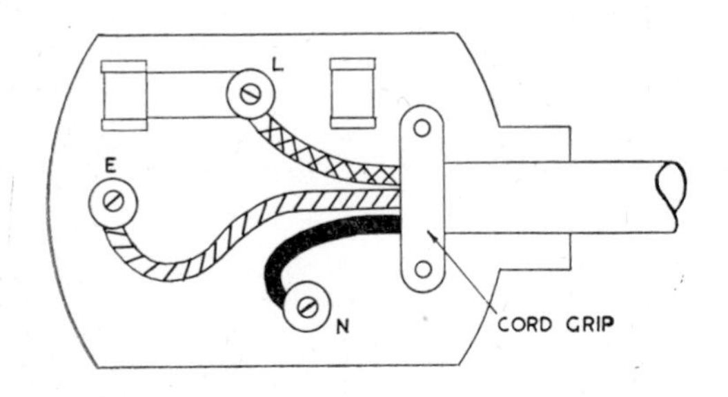

FIG. 1.6 Exercise 1.3

Procedure

1. Remove outer covering of the flexible cord for a suitable distance.
2. Separate the three cores and measure off a suitable length for each by referring to the appropriate terminal in the plug top.
3. Remove the insulation from the end of each core and connect to the terminals.

The following points must be carefully observed:

(*a*) That the red core is connected to the terminal marked L, the black core to the terminal marked N, and the green core to the terminal marked E. (Fig. 1.6.)
(*b*) The flexible cord is securely held by the cord grip at the point where it enters the plug top.
(*c*) That every strand of each core is securely held by its appropriate terminal.

Answer the following questions

1. Why is it important that each particular coloured core is connected to the appropriate terminal?
2. What could be the result of connecting the green core to the terminal marked L?

Exercise No. 1.4 — To terminate lead-covered V.R.I. cable in a switch.

Materials

3/·029 Lead-sheathed V.R.I. cable.
1 Hard wood round block.
1 5A Single-way lighting switch.

Diagrams

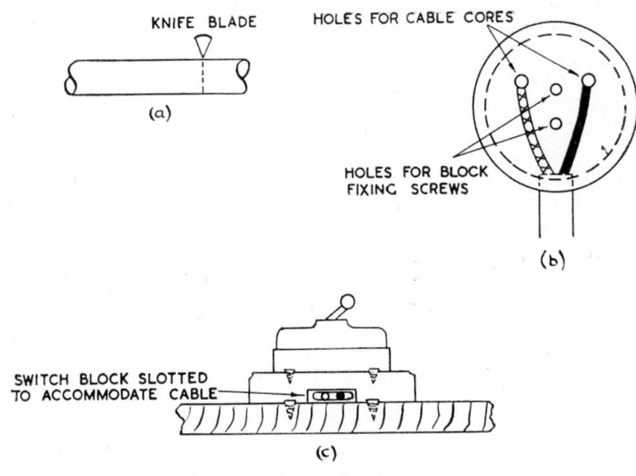

FIG. 1.7 (a–c) Exercise 1.4

(a) Cable sheath scored all round; (b) Drilling switch block; (c) Switch and block fixed to board.

Procedure

1. Score the cable sheath all round taking care not to cut the lead completely through. Then bend the cable backwards and forwards until the lead sheath fractures at the scored mark. The short end of the lead sheath can then be pulled off.
2. Remove the switch cover, place the switch on the centre of the block and carefully mark the positions for the cable cores and the switch fixing screws. Remove the switch and drill holes in the block to accommodate the block fixing screws and the cable cores in the positions indicated by Fig. 1.7 (b).

3. Slot the switch block in the appropriate position sufficient only to avoid the cable being 'pinched' by the block. (Fig. 1.7 (c).)
4. Lay the cable in the slot and push a sufficient length of the cable's cores through the corresponding holes.
5. Fix the switch block to the board, terminate the cable cores in the switch terminals and fix the switch to the block.

Answer the following questions

1. Why are two screws used to fix the switch block to the board?
2. Why is it bad practice to use only one screw to fix the switch to the block and the block to the board?

EXERCISE No. 1.5 — To terminate T.R.S. cable in a joint box.

Materials

3/·029 Two-core T.R.S. cable.
1 Bakelite joint box.

Diagram

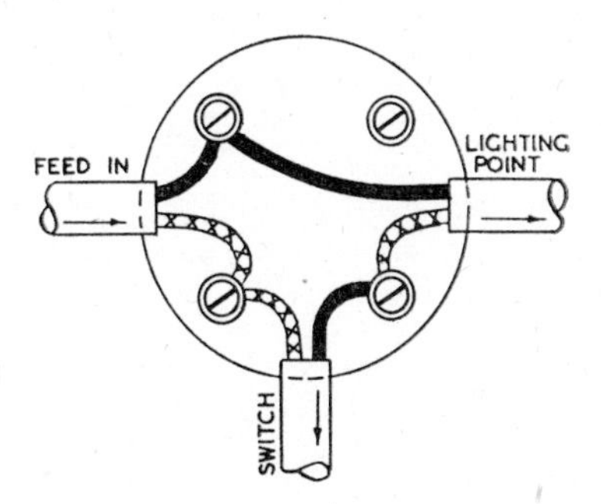

FIG. 1.8 Exercise 1.5

Procedure

1. The joint box must be prepared by breaking away the bakelite at the required cable entries, using a pair of snipe-nosed pliers. Great care is needed with this operation to ensure a neat result.
2. Divide the cable into three equal lengths and remove a suitable length of the tough rubber sheath from the end of each.
3. Cut cores to a suitable length by checking against the joint box, and remove sufficient insulation from the ends of each core to suit the terminals. Tightly twist the cores (in the direction of the lay of the cable), insert into the appropriate terminals, as shown in Fig. 1.8, and tighten the grub screws.

Points to be observed

(*a*) Do not nick the insulation of the conductors when removing the outer rubber covering.
(*b*) Do not nick the conductors when removing the insulation from them.
(*c*) The cores should if anything be cut rather too long and the excess 'lost' in the joint box (see sketch). On no account should the cores be stretched or sharply bent.

(*d*) The tough rubber sheath of each cable must be taken well into the joint box.

Answer the following questions

1. Define the terms joint box and junction box.
2. Why is it considered good practice to allow a small amount of extra core length in the joint box?

EXERCISE No. 1.6 — To terminate lead-sheathed V.R.I. cable in a joint box.

Materials

3/·029 Two-core lead-sheathed V.R.I. cable.
1 Metal joint box (tee-box type).
1 Three-way porcelain shrouded connector.

Diagram

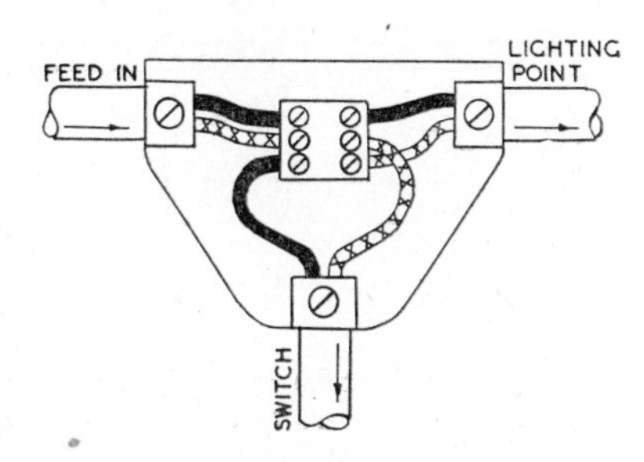

FIG. 1.9 Exercise 1.6

Procedure

1. Divide the cable into three equal lengths and remove a suitable length of the lead sheath from the end of each. A good method of stripping the lead sheath is to score it carefully all round taking care not to cut the lead completely through. The end of the cable is now bent backwards and forwards until the lead sheath fractures at the scored mark after which the short end of lead sheath can be pulled off.
2. Remove the insulation from the ends of the cable cores and connect to the porcelain connector as shown in Fig. 1.9.

Points to be observed

(*a*) Ensure that the lead sheaths are firmly held by the bonding clamps.
(*b*) Ensure that the insulation of the cores is not damaged when removing the lead sheath.
(*c*) Do not nick the conductors when stripping the insulation from them.

Answer the following questions

1. Explain, with the aid of sketches, how lead-sheathed cables may be used in conjunction with metal conduits.
2. Why must special care be taken in handling lead-sheathed cable?

EXERCISE No. 1.7 — Married joints in 3/·029 V.R.I. taped and braided cable.

Materials

3/·029 V.R.I. taped and braided cable.

Diagrams

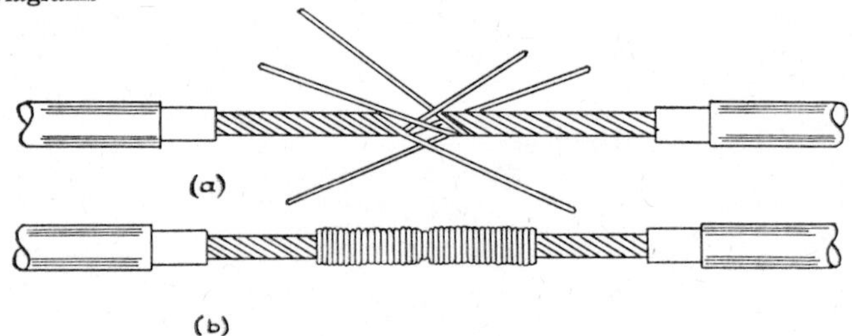

FIG. 1.10 (a–b) Exercise 1.7
(a) Interleaving the splayed strands; (b) Completed joint.

Procedure

1. Divide the cable into two equal lengths and remove tape, braid and insulation from one end of each length of cable to leave 3 in. of exposed conductor.
2. Further remove the tape and braid to expose $\frac{1}{2}$ in. of insulation.
3. Ensure that there are no particles of rubber adhering to bared tinned copper strands.
4. Twist the conductors tightly together in the direction of the lay of the cable for 1 in. leaving 2 in. splayed out.
5. Interleave the splayed out strands of the cables as shown in Fig. 1.10 (a).
6. Hold down the strands of the right hand cable along the left hand cable, and wrap the three strands of the left hand cable neatly around the right hand cable half a turn at a time.
7. Wrap the strands of the right hand cable neatly around the left hand cable. Tighten the joint using pliers. (Fig. 1.10 (b).)

Answer the following questions

1. What do the I.E.E. Regulations stipulate with respect to accessibility of joints?
2. Why is it important to ensure that there are no particles of rubber adhering to the strands?

EXERCISE No. 1.8

Tee joint in 7/·029 V.R.I. taped and braided cable.

Materials

7/·029 Taped and braided V.R.I. cable.

Diagram

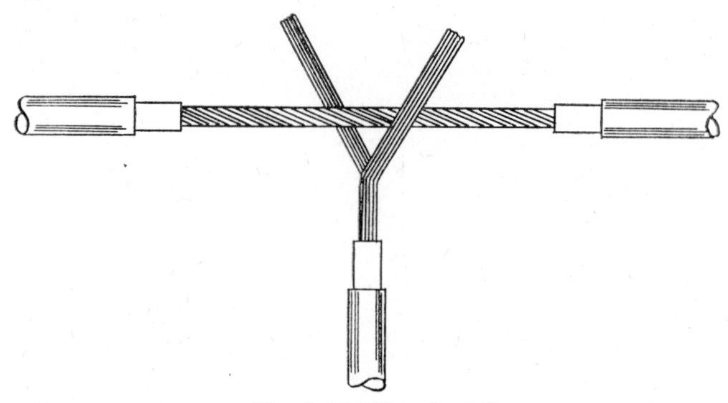

FIG. 1.11 Exercise 1.8

Procedure

1. Divide the cable into two lengths, one 9 in. long and the other 1 ft. 3 in.
2. Remove tape, braid and insulation for 3 in. from the centre of the longer length and then remove the tape and braid for a further $\frac{1}{2}$ in. from the ends of the insulation.
3. Remove tape, braid and insulation for 3 in. from the end of the shorter length and then remove the tape and braid for a further $\frac{1}{2}$ in. from the end of the insulation and clean the strands.
4. Twist the conductors of the shorter length tightly together in the direction of the lay of the cable for 1 in. and then divide the strands.
5. Butt the fork of the tee piece (shorter length) against the centre of the through piece (longer length) and carefully wrap three strands of the tee piece around one side of the through piece. The strands should be wrapped neatly, half a turn at a time. Wrap the remaining four strands around the other side of the through piece, tighten the joint using pliers.

Answer the following questions

1. Where might this type of joint be used?
2. Why should a joint be mechanically and electrically sound before it is soldered?

EXERCISE No. 1.9 **To solder a married joint.**

Materials

The married joint made in a previous exercise.
Solder and Flux.

Diagram

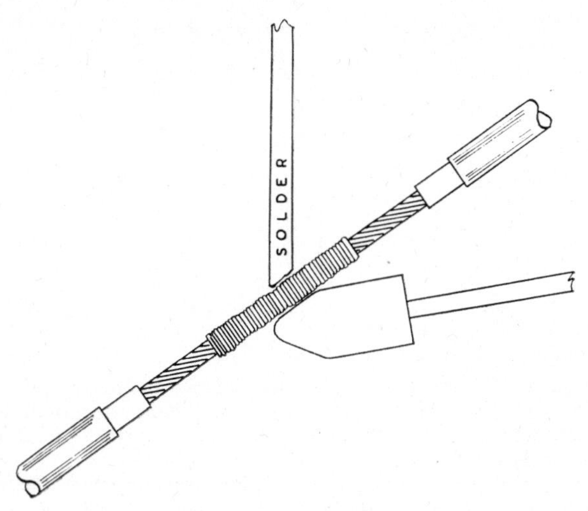

FIG. 1.12 Exercise 1.9

Procedure

1. Apply a light film of flux to the twisted joint.
2. Heat the soldering iron until solder flows freely on its surface.
3. Quickly clean the bit of the iron with a file and tin it by applying a little flux and solder.
4. Hold the soldering iron firmly against the joint and apply a stick of solder, until solder runs freely all over the joint. When certain that the solder has completely penetrated the joint, quickly wipe any surplus solder from the joint using a clean rag.
5. After the joint has cooled remove any sharp points or blobs of solder using a file.

Points to be observed

(*a*) The joint must be electrically and mechanically sound *before* soldering.
(*b*) Neat and careful workmanship is essential.
(*c*) The soldering process must be performed as quickly as possible to avoid an undue quantity of heat being conducted along the wires to the rubber insulation.

Answer the following questions

1. Why is flux necessary?
2. Why should a married joint be soldered?

Exercise No. 1.10 — To solder a cable socket to a length of 19/·044 V.R.I. taped and braided cable.

Materials

19/·044 Taped and braided V.R.I. cable.
1 60A Cable socket.

Diagrams

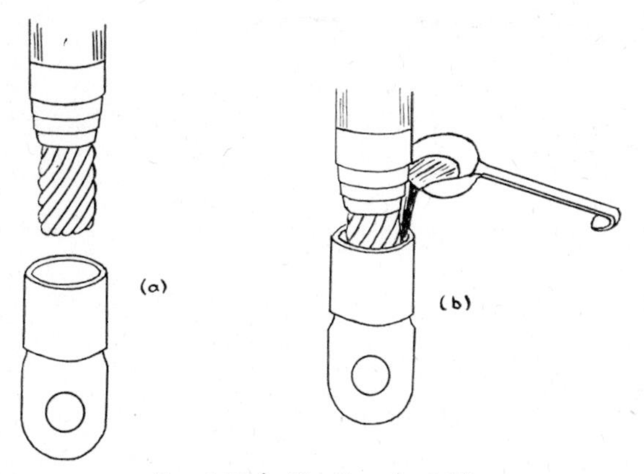

FIG. 1.13 (a–b) Exercise 1.10
(a) Inserting prepared cable into socket which is already $\frac{3}{4}$ filled with solder; (b) Making up level of solder.

Procedure

1. Remove tape, braid and insulation from one end of the cable leaving sufficient exposed conductor to touch the bottom of the cable socket leaving $\frac{1}{2}$ in. clearance between the rim of the cable socket and the end of the insulation.
2. Remove the tape and braid a further $\frac{1}{2}$ in. from the end so that the insulation is exposed.
3. Ensure that there are no particles of rubber adhering to the copper strands.
4. Twist the conductors tightly together in the direction of the lay of the cable.
5. Bind the exposed insulation with linen tape.
6. Ensure that the inside of the socket is clean and apply a film of flux.
7. Melt the solder and skim any impurities from the surface and warm the ladle.

8. Fill the socket with molten solder and tip it out quickly. Inspect to see if the inside of the socket is completely tinned, if not repeat the process.
9. Apply a film of flux to the inside of the socket and to the exposed conductors of the cable.
10. Three-quarters fill the cable socket with molten solder and then push the exposed cable conductors quickly and carefully into the socket.
11. Ensure that the cable is held vertically in the centre of the socket and is not disturbed until the solder is set.
12. If on setting the solder shrinks too far below the rim of the socket pour in molten solder until the correct level is obtained.
13. Remove any surplus solder from the outside of the cable socket.
14. Remove linen tape and cut away damaged insulation.

Taping

The insulation of the exposed conductors should be made good by neatly wrapping pure rubber tape around them to the same thickness as the original insulation. A binding of black adhesive tape may be used to secure the outside end of the rubber tape.

Answer the following questions

1. When is it advisable to use a cable socket?
2. Why is it good practice to apply a temporary binding of linen tape over the exposed insulation?

CHAPTER 2

Electric Circuits

2.1 Series and Parallel Circuits

(*a*) An electric circuit is defined in the I.E.E. Regulations as 'an arrangement of conductors for the purpose of carrying current'.

Before an electric current can flow in a circuit two conditions must be fulfilled:

(i) There must be a source of electro-motive-force (e.m.f.) to overcome the resistance of the circuit and so force the current around it.

(ii) There must be a complete path of conducting materials through which the current can flow.

(*b*) In a series circuit the connections are such that the electric current flows through each part of the circuit in turn. (Fig. 2.1.)

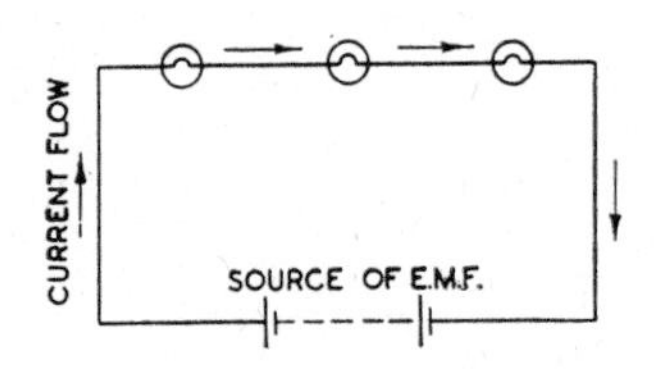

FIG. 2.1 Lamps connected in series

Each component in a series circuit receives the same amount of current but the supply voltage is shared between the components. This type of circuit is not suitable for general use as:

(i) It is easier to provide a supply at constant voltage than to provide a constant current supply.

(ii) It is difficult to provide suitable switching arrangements to control individual parts of a circuit.

There are, however, some circumstances in which series circuits can be used to advantage, for example, decorative lighting circuits using a number of identical low voltage lamps, and battery charging arrangements where a number of the cells to be charged can be connected in series.

(*c*) In the parallel circuit the connections are such that the same voltage is applied to each component. (Fig. 2.2.)

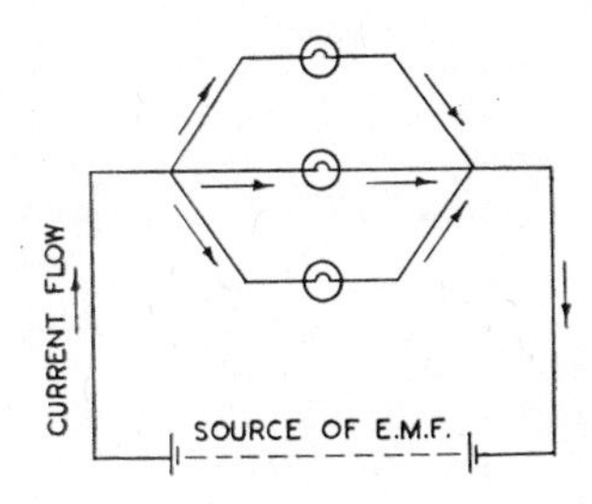

FIG. 2.2 Lamps connected in parallel

Each component in a parallel circuit receives the same voltage but the total current supplied is shared between the components. This type of circuit is widely used as:

(i) It is easy to arrange a supply at constant voltage.

(ii) It is easy to switch off individual parts of a circuit without affecting the operation of other apparatus connected to the circuit.

The e.m.f. needed to operate a circuit may be provided by a cell or battery of cells. Some installations use their own generating plant but for the majority of installations the supply is derived from the mains.

2.2 Typical Voltage Ranges

The table below gives the designations and ranges of the voltages which may be encountered in electrical work, together with a brief list of the applications of each voltage range.

Designation	Voltage Ranges	Some Applications
Extra low voltage (E.L.V.).	30V r.m.s. a.c. or 50V d.c. or less	Special lighting circuits, bell circuits, etc.
Low voltage (L.V.).	250V or less	Domestic installations and small industrial installations.
Medium voltage (M.V.).	250V to 650V	Larger industrial and commercial installations.
High voltage (H.V.).	650V to 3,000V	Local distribution by supply authorities and distribution in large factories.
Extra high voltage (E.H.V.).	Over, 3,000V	Large-scale distribution networks.

2.3 Switch Control of Lighting Circuits

Lighting circuits are generally controlled using:

(i) Single pole, one-way switches.
(ii) Single pole, two-way switches.
(iii) Intermediate switches.

(a) *One-way switch circuit.* (Fig. 2.3.)

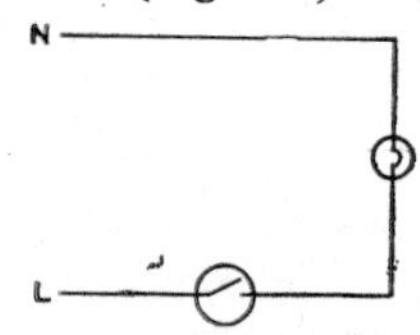

FIG. 2.3 One-way switch circuit

This circuit provides 'on/off' control. When the switch is closed the lamp is on, when the switch is open the lamp is off.

(b) *Two-way switch circuit.* (Fig. 2.4.)

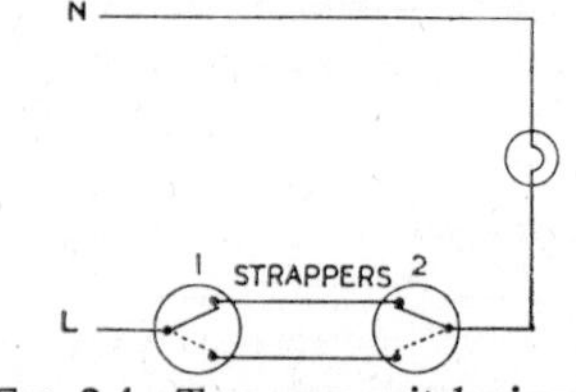

FIG. 2.4 Two-way switch circuit

This circuit provides a way of controlling a lamp from two independent switch positions. The following table shows how the circuit operates.

Operation of Two-way Switch Circuit

SWITCH POSITION		CIRCUIT CONDITION	LAMP
Switch 1	*Switch 2*		
Up	Up	Circuit complete	On
Up	Down	Circuit broken	Off
Down	Up	Circuit broken	Off
Down	Down	Circuit complete	On

This circuit is often used on stairways and halls.

(c) *Two-way and intermediate switch circuit.* (Fig. 2.5.)

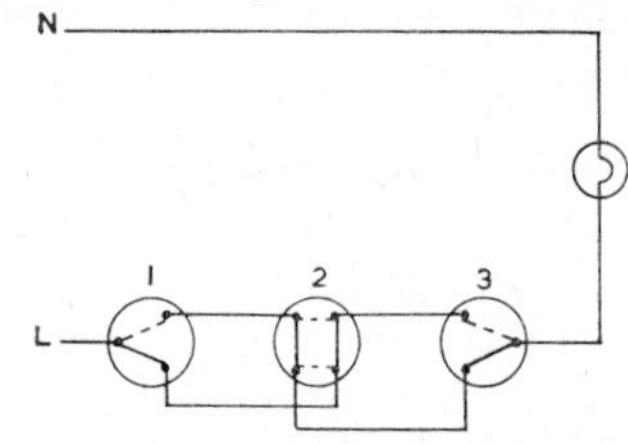

FIG. 2.5 Intermediate switch circuit

This circuit provides a way of controlling a lamp from three independent switch positions. The internal connections made by the intermediate switch are indicated in the diagram by: – – UP POSITION; —— DOWN POSITION. The following table shows how the circuit operates.

Operation of Two-way and Intermediate Switch Circuit

SWITCH POSITION			CIRCUIT CONDITION	LAMP
Switch 1	*Switch 2*	*Switch 3*		
Up	Up	Up	Circuit complete	On
Up	Up	Down	Circuit broken	Off
Up	Down	Up	Circuit broken	Off
Up	Down	Down	Circuit complete	On
Down	Up	Up	Circuit broken	Off
Down	Up	Down	Circuit complete	On
Down	Down	Up	Circuit complete	On
Down	Down	Down	Circuit broken	Off

2.4 Polarity

(a) *Switches*. It cannot be too strongly emphasized that, in the interests of safety, switches must be connected in the 'live' (phase) conductor. This is to ensure that when the switch is open the circuit is disconnected from the live side of the supply, and so it becomes possible to replace lamps or do other work on the circuit without risk of electric shock. If switches are incorrectly connected in the neutral conductor the circuit can still be switched on and off, but even though the lamps are 'off' the wiring and fittings remain connected to the live conductor and are, therefore, 'live' themselves and there is considerable risk of shock to anyone attempting to work on the circuit.

(b) *Plugs and sockets*. For similar reasons care must be taken when connecting socket outlets and plug tops to ensure that:

(i) The 'live' conductor is connected to the terminal marked 'L'.

(ii) The 'neutral' conductor is connected to the terminal marked 'N'.

(iii) The 'earth' conductor is connected to the terminal marked 'E'.

(c) *Lamp holders*. Edison screw lamp holders must have the centre contact connected to the 'live' conductor and the outer (screwed) contact connected to the 'neutral' conductor to reduce the danger of shock should the fittings be touched when the lamps are on.

2.5 Sub-circuits

(*a*) The supply authorities are responsible for providing a supply to suitable terminals on the consumer's premises, the electrical contractor is concerned with the installation of the circuits which start from the supply authority's 'live' terminal and finish at the 'neutral' terminal.

As the circuit of a completed installation can be complex, offering many paths to the passage of current, it is usually necessary to divide the installation into numerous 'sub-circuits'.

(*b*) 'Table A' of the I.E.E. Regulations which is reproduced below gives the general sequence of the necessary equipment for controlling the supply entering a consumer's premises. In some cases 'final sub-circuits' are supplied directly from the 'consumer's circuit fuses' mentioned in the table but in larger installations the initial consumer's fusing and control arrangements may be used to supply further distribution boards to which the final sub-circuits are connected.

TABLE A (*I.E.E. Regulations*)
Sequence of supply controls

	Service cable and sealing box (if any); Service fuse(s), and neutral link (if any); watt-hour meter;		(Supply undertaking's equipment)
linked switch.	or linked overload circuit-breaker functioning also as earth-leakage device, where the impedance of earth-leakage path is sufficiently low.	or linked circuit-breaker with independent overload and earth-leakage coils.	or linked switch*
consumer's main fuse functioning as excess-current and earth-leakage device, where the impedance of earth-leakage path is sufficiently low.			
	Consumer's circuit fuses		

Note. – Where a circuit-breaker of insufficient breaking capacity for short-circuit faults is installed, fuse(s) of adequate breaking capacity may be required in addition.

* It may be necessary for the switch and fuses to be of the 'all insulated' pattern.

(*c*) Fig. 2.6 shows a typical sequence of supply controls. The incoming supply enters via the supply authority's equipment to a main switch of the 'linked type' (i.e. a switch which breaks both poles of the supply). From the main switch the supply is fed to a distribution board containing fuses or other suitable overload protection equipment. Some final sub-circuits are connected directly to this distribution board, while a 'sub-main' is used to supply a further 'sub-distribution board' to which other final sub-circuits are connected.

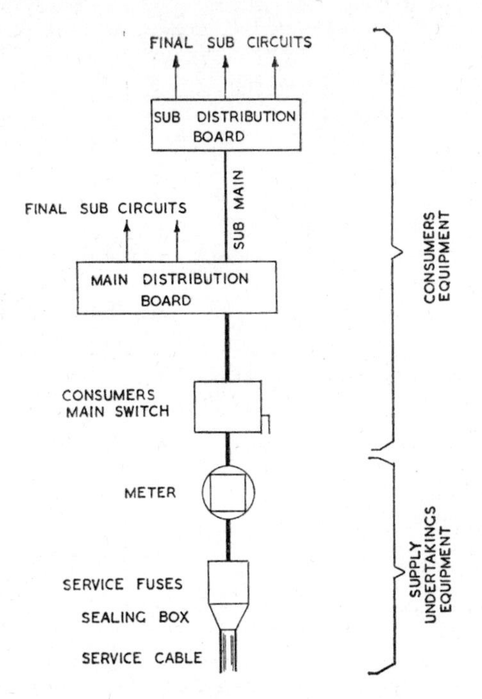

FIG. 2.6 Sequence of supply controls

(*d*) A 'final sub-circuit' is defined as an outgoing circuit connected to a distribution board and intended to supply electrical energy direct to current using apparatus. Final sub-circuits may be broadly divided into two groups, those in which the current demand is less than 15A and those in which the current demand exceeds 15A.

(*e*) The requirements of the I.E.E. Regulations for final sub-circuits of rating not exceeding 15A are:

(i) The number of points which may be supplied is limited by their aggregate demand as determined from Table B below.

(ii) There must be no allowance for diversity (para. 2.7 (c)).

(iii) The current rating of the cable must not be exceeded (para. 2.7 (b)).

TABLE B (*I.E.E. Regulations*)
Assumed current demands of points

Point or appliance	*Current demand to be assumed*
(*a*) 15-ampere socket-outlet.	15 amperes.
(*b*) 13-ampere socket-outlet.	See Table C.
(*c*) 5-ampere socket-outlet.	5 amperes.
(*d*) 2-ampere socket-outlet.	At least $\frac{1}{2}$ ampere.
(*e*) Lamp holder.	Current equivalent to the connected load, with minimum of 100 watts.
(*f*) Electric clock, or electric shaver supply unit complying with B.S.3052.	May be neglected.
(*g*) Other fixed or free-standing appliance.	British Standard rated current, or normal current (see also Regulation 114(B)).

In the interest of good planning it is not always advisable to install the maximum permitted number of outlets for any final sub-circuit. For example, it is permissible to use a lighting circuit rated at 15A; this means (assuming a supply voltage of 250V) that the circuit can supply approximately 37 outlets; however, in order to do this it would be necessary to use 15A rated cables and flexibles throughout the circuit, and furthermore should the fuse blow a very large number of lamps are put out of action at the same time. In most cases it is better to limit the rating of each lighting sub-circuit to 5A, i.e. approximately 12 outlets, and in domestic installations a natural division can often be made by using a separate sub-circuit for each floor of the building. 15A rated lighting circuits can only be used with advantage where very large numbers of lighting points are required, such as in industrial installations.

(*f*) The requirements of the I.E.E. Regulations for final sub-circuits exceeding 15A are that these sub-circuits shall not supply more than one point, with the following exceptions:

(i) Final sub-circuits of ratings in the range 15 to 30A may be used to supply a number of 13A rated socket outlets for use with fused plugs provided that the circumstances are as stated in section 2.6 below.

(ii) A 30A rated final sub-circuit may supply a 30A cooker control unit which also includes one socket outlet.

2.6 Use of Fused Plugs and Socket Outlets

As a fused plug provides local protection for the appliance connected to it there are several advantages to be gained by using socket outlets designed to accept only this type of plug. Such socket outlets are normally rated at 13A and so can supply apparatus consuming up to 3kW. Several such socket outlets can be supplied by a single sub-circuit rated in the range 15 to 30A provided that the circumstances are as listed below:

(*a*) A non-ring circuit, wired using 7·029 cable and protected by a 20A rated fuse can be used to supply two 13A socket outlets. The same size cable and fuse can be used to supply three 13A socket outlets provided that all the socket outlets are installed in one room of a house or flat, which is not a kitchen.

(*b*) A non-ring circuit, wired using 7/·036 cable and protected by a 30A rated fuse can be used to supply up to six 13A socket outlets.

(*c*) A ring circuit is a final sub-circuit in which the current carrying and earth continuity conductors are connected in the form of loops, with both ends of each loop connected to a single way in a distribution fuse board or its equivalent. Such a circuit may be wired using 7/·029 cable protected by a 30A rated fuse, and used to supply up to ten 13A rated socket outlets.

(*d*) If ring circuits are installed in a domestic premises on the basis of one ring circuit for every 1,000 ft.^{2} of floor area, then an unlimited number of 13A socket outlets may be supplied. When advantage is taken of this, it is good practice to arrange that each ring supplies approximately the same number of socket outlets.

(*e*) Fixed appliances of not more than 13A rating may be connected to a ring circuit, provided they are connected either by means of a fused plug and socket outlet or by using a local fuse or circuit-breaker to provide the necessary protection. Each fixed appliance counts as one socket outlet when assessing the number of socket outlets supplied by the ring.

(*f*) It is permissible to connect spurs, or branch cables of cross-sectional-area not less than that of the conductors forming the ring, to a ring circuit. No spur may supply more than two socket outlets or one fixed appliance and not more than half of the points installed may be fed by spurs.

2.7 Planning an Installation

(*a*) In order to choose a suitable wiring system for any installation the electrical contractor must be familiar with the merits of each of the large number of alternative systems which are available. Chapters 3 and 4 of this book describe the principal methods in use at the present time and also discuss the factors which must be taken into account when choosing a wiring system. When selecting the size of cable to be used in a particular circuit there are two main factors to be considered; the cable must be able to carry the maximum current liable to flow in the circuit without undue heating; and the voltage drop caused by the resistance of the cable must not be excessive.

(*b*) The current rating of a conductor is the maximum current that it can carry continuously without undue heating. As the temperature rise of a conductor depends on both the amount of current flowing and on the situation in which it is installed (i.e. on the rate at which the heat can escape) the current rating for a particular size conductor can vary. The current ratings of various types of cables installed in particular conditions are listed in I.E.E. Tables 12 to 27. In certain situations such as where several cables are enclosed in the same enclosure or where conditions are warmer than usual the current rating given in the tables must be adjusted by multiplying by an appropriate rating factor; these rating factors are also listed in the tables mentioned above. In many cases the rating factor is determined by considering the ambient air temperature, that is the temperature of the air in the immediate vicinity of the cables when no current is flowing, allowance for the normal groupings of cables being allowed for in various columns of the tables. Where larger than normal numbers of cables are grouped in an enclosure the combined rating factor is found by multiplying the rating factor for temperature by the rating factor for grouping. When determining the correct cable size for use in a situation where a rating factor is applicable, one method is to modify the ratings given in the I.E.E. Tables by multiplying each by the rating factor, and then to choose the cable size whose modified rating is just higher than the current to be carried. It is, however, more convenient to divide the actual current to be carried by the rating factor, so obtaining a figure corresponding to the required current rating under normal conditions, the correct size of cable can then be chosen directly from the tables. The following problem gives an example in the use of rating factors.

Example No. 1

Two lighting circuits each supplying a load of 2kW at 240 volts are to be installed. Circuit A is situated in a drawing office where the temperature is unlikely to exceed 90° F. Circuit B is situated in a heat treatment shop where the temperature in the vicinity of the cables is liable to reach 113° F. Both circuits are to be wired using single core P.V.C. insulated cables in conduit, determine a suitable cable size for each installation.

Solution

Circuit A

$$\text{Actual current in conductors} = \frac{2{,}000}{240} = 8{\cdot}3\text{A.}$$

For temperature not exceeding 90° F. no rating factor is needed.

From I.E.E. Table 12, 3/·029 cable is rated at 10A this is the nearest current rating above 8·3A so this is the size of cable required.

Circuit B

For ambient temperature of 113° F. the rating factor = 0·47.

$$\text{The actual current in the conductors} = \frac{2{,}000}{240} = 8{\cdot}3\text{A.}$$

Because of the rating factor required to allow for the high ambient temperature, the cable must be capable of carrying a current of $\frac{8{\cdot}3}{0{\cdot}47} = 17{\cdot}7$A under normal conditions.

By consulting Table 12 of the I.E.E. Regulations it is found that 7/·029 cable (rated at 20A for normal conditions) is the size of cable required.

As a check on this solution, note that the actual rating of 7/·029 cable under the specified conditions is 20 × 0·47 = 9·4A. This is just higher than the required 8·3A so the cable will be satisfactory.

(*c*) In many cases it is unlikely that all the apparatus connected to a circuit will be switched on at the same time, thus the maximum current that the conductors will actually have to carry will be less than the total possible current which would flow if all the apparatus were switched on simultaneously.

$$\text{DIVERSITY FACTOR} = \frac{\text{Actual maximum current}}{\text{Total current required for all apparatus combined}}$$

The estimation of diversity factors requires great experience. A helpful guide is Table 1 of the I.E.E. Regulations which lists the factors which apply for a variety of common situations. When applying a diversity factor the current rating of the cable required is found by multiplying the full load current of all the apparatus connected by the diversity factor. Diversity factors may be applied to all circuits other than final sub-circuits and to certain final sub-circuits, for example a ring circuit supplying 13A socket outlets.

(*d*) When selecting cables for a particular installation the voltage drop in the cable must be considered as well as the current rating. The I.E.E. Regulations state that under normal conditions of service the voltage drop when the cable is carrying maximum current shall not exceed 1 volt plus 2% of the declared nominal voltage of the supply. Two exceptions are allowed:

(i) If some form of automatic regulator is installed to maintain a constant voltage at the consumer's terminals then a voltage drop of 5% of the declared voltage is permitted.

(ii) In motor circuits a voltage drop at full load current of up to 7·5% of the declared voltage is permitted provided that satisfactory starting can be obtained.

(It should be noted that in general the starting current of a motor is greater than its normal full load current thus giving a greater voltage drop.)

(*e*) The calculation of voltage drops can be simplified by using the figures for the length of run for one volt drop at rated current available in the I.E.E. Tables. When using these figures it must be remembered that:

(i) Voltage drop is directly proportional to the length of run.
(ii) Voltage drop is directly proportional to the current flowing.

$$\text{VOLTAGE DROP} = \frac{\text{Actual length of run}}{\text{Length of run for 1 volt drop}} \times \frac{\text{actual current}}{\text{rated current}}$$

Example No. 2

The I.E.E. Tables state that a 7/·029 single-core P.V.C. cable has a rated current of 20A and the length of run for 1 volt drop at rated current is 12 ft. Find the voltage drop in a 30 ft. length of run when the current is 14A.

Solution

$$\text{Voltage drop} = \frac{30}{12} \times \frac{14}{20}$$

$$= 1{\cdot}75\text{V.}$$

(*f*) When a rating factor has been used to modify the current rating of a cable, the quoted figure for length of run for 1 volt drop should also be modified by dividing it by the rating factor. In Example 2 the 7/·029 cable used had a current rating of 20A and the length of run for 1 volt drop was 12 ft. If a rating factor of 0·86 had to be applied these figures would become:

$$\text{Current rating} = 20 \times 0{\cdot}86 = 17{\cdot}2\text{A.}$$

$$\text{Length of run for 1 volt drop} = 12 \div 0{\cdot}86 = 16{\cdot}28 \text{ ft.}$$

However, as the student can check by working Example 2 using these modified figures, the voltage drop caused by a current of 14A flowing through a 30 ft. run is still 1·75V. Thus when calculating voltage drops it is usually easier to use the values of current rating and length of run as quoted in the tables rather than to modify *each* figure by the appropriate rating factor. A typical problem involving the consideration of the various factors discussed above is given below.

Example No. 3

Select a suitable cable size for a 240V single-phase sub-main serving a lighting load totalling 12kW. Length of run 46 ft., average temperature 77° F., diversity factor 80%. The circuit is to be wired using V.R.I. taped and braided cables in conduit:

Solution

$$\text{Total current for all lamps combined} = \frac{12{,}000}{240} = 50\text{A.}$$

Allowing for diversity factor:

$$\text{Actual maximum current} = \frac{50 \times 80}{100} = 40\text{A.}$$

Rating factor for 77° F. = 1·13.

$$\text{Equivalent current under normal conditions} = \frac{40}{1{\cdot}13} = 35{\cdot}4\text{A.}$$

The nearest cable size with a rating higher than this is 7/·044 cable rated at 36A under normal conditions.

Before deciding to use this size cable, the voltage drop must be checked.

From the tables the rated current is 36A and the length of run for volt drop at rated current is 16 ft.

$$\text{Voltage drop} = \frac{40}{36} \times \frac{46}{16} = 3{\cdot}19\text{V.}$$

(Note that in this calculation the actual maximum current of 40A must be used, not the equivalent current of 35·4A under normal conditions which was used to make the preliminary selection of cable size. Note also that the figures for rated current and length of run for 1 volt drop are taken directly from the tables.)

$$\text{Permitted voltage drop} = 1 + \frac{240 \times 2}{100} = 5{\cdot}8\text{V}.$$

As the calculated voltage drop is less than the permitted value the 7/·044 cable will be suitable for this application.

(*g*) When choosing the route to be followed by the wiring in a building the following factors must be taken into consideration as far as possible:

(i) Cables should not be located in positions where they are subject to the risk of mechanical damage, or are liable to deteriorate because of the adverse effects of vibration, moisture, heat, corrosive environments, etc.

(ii) It is an advantage if the runs are easily accessible both for installation and maintenance, but at the same time the wiring should be as unobtrusive as possible.

(iii) The routes chosen should run as directly as possible, so avoiding the use of an excessive quantity of material and keeping voltage drops to a minimum.

The relative importance of these factors varies so much from one installation to another, that no hard and fast rules can be laid down. A good knowledge of the properties of the various types of wiring system allied with experience of the problems that arise in various types of building provides the best guide to the selection of suitable runs.

EXERCISES

The exercises which conclude this chapter are designed to illustrate the connections of some typical lighting circuits. All the exercises are wired using single-core cables and standard electrical fittings. It must be emphasized that these exercises are not intended to represent practical wiring systems. These are dealt with in Chapters 3 and 4. It should be noted that in the diagrams relating to the following exercises a fine line has been used to represent red and a bolder line to represent black coloured cable.

EXERCISE No. 2.1 **Two lighting points each separately controlled using the loop-in wiring system.**

Materials

3/·029 V.R.I. Red.
3/·029 V.R.I. Black.
14/·0076 Flexible cord.
2 5A one-way lighting switches.
2 Two-plate ceiling roses.
2 B.C. lamp holders.
4 Hard wood round blocks.

Diagrams

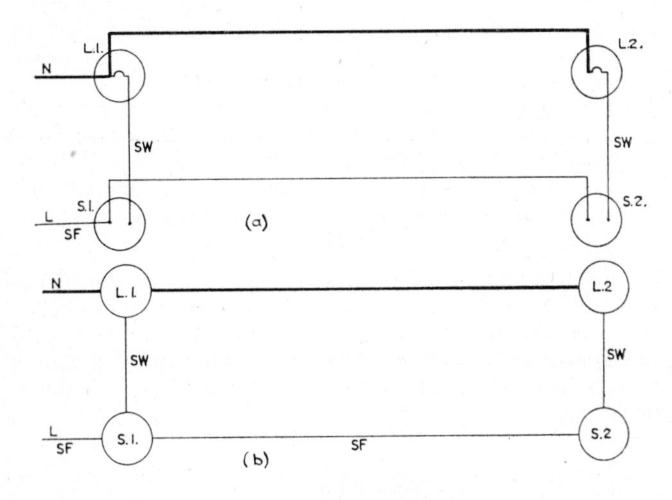

FIG. 2.7 (a–b) Exercise 2.1
(a) Connections; (b) Layout.

Procedure

Connect the circuit as shown in Figs. 2.7 (a) and (b).

Answer the following questions

1. Explain what is meant by the 'loop-in' wiring system.
2. What are the main advantages of this wiring system?

EXERCISE
No. 2.2

Two lighting points each separately controlled using three-plate ceiling roses.

Materials

3/·029 V.R.I. Red.
3/·029 V.R.I. Black.
14/·0076 Flexible cord.
2 5A one-way lighting switches.
2 Three-plate ceiling roses.
2 B.C. lamp holders.
4 Hard wood round blocks.

Diagrams

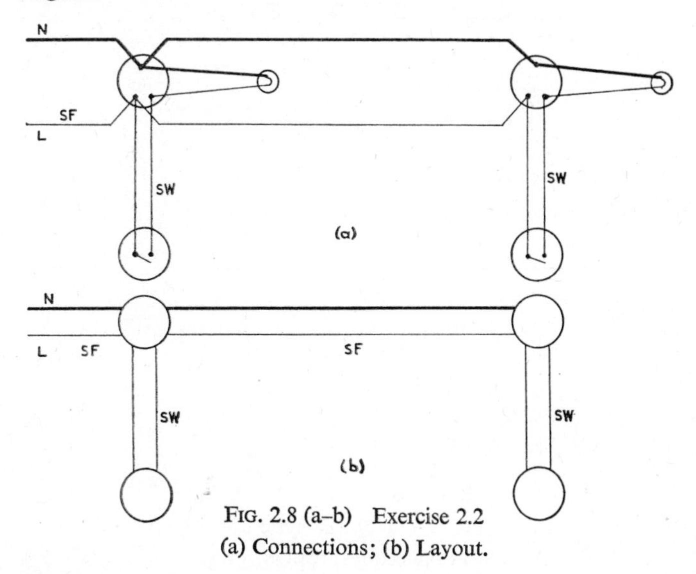

FIG. 2.8 (a–b) Exercise 2.2
(a) Connections; (b) Layout.

Procedure

Connect the circuit as shown in Fig. 2.8 (a) and (b).

Answer the following questions

1. Why should switch feeds and switch wires be wired in red-coloured cables?
2. What are the main advantages of using three-plate ceiling roses?

EXERCISE NO. 2.3 **Five lighting points controlled by three separate switches.**

Materials

3/·029 V.R.I. Red.
3/·029 V.R.I. Black.
5 B.C. Batten holders.
3 5A one-way lighting switches.
5 Hard wood round blocks.
1 Hard wood rectangular block.

Diagrams

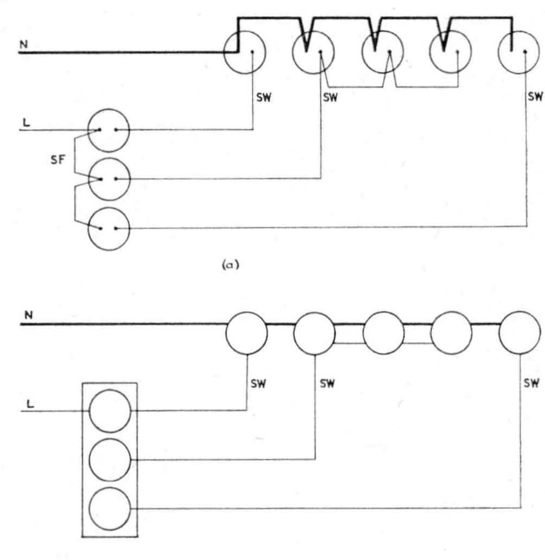

FIG. 2.9 (a–b) Exercise 2.3
(a) Connections; (b) Layout.

Procedure

Connect the circuit as shown in Fig. 2.9 (a) and (b).

Answer the following questions

1. What do you understand by the term 'final sub-circuit'?
2. Given that the supply is at 200V, how many lighting points may be supplied by a 5A rated final sub-circuit? (Each lighting point is deemed to consume 100 watts.)

EXERCISE No. 2.4 — Two-way control of a lighting point.

Materials

3/·029 V.R.I. Red.
3/·029 V.R.I. Black.
14/·0076 Flexible cord.
2 5A two-way switches.
1 Two-plate ceiling rose.
1 B.C. lamp holder.
3 Hard wood round blocks.

Diagrams

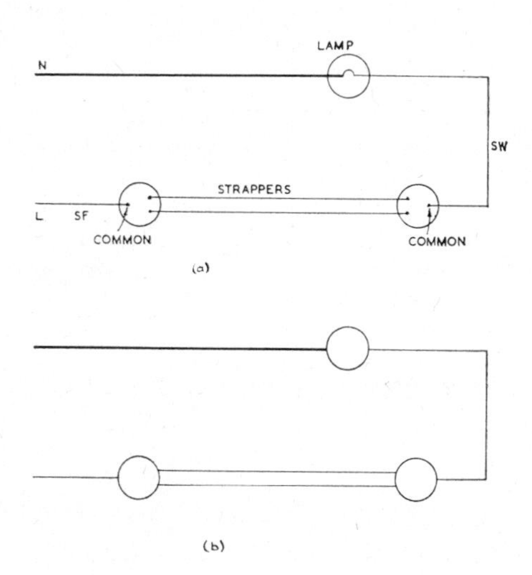

FIG. 2.10 (a–b) Exercise 2.4
(a) Connections; (b) Layout.

Procedure

Connect the circuit as shown in Fig. 2.10 (a) and (b).

Answer the following questions

With the aid of neatly drawn circuit diagrams show the path taken by the current for each possible position of the switches, and so explain the operation of the circuit.

EXERCISE No. 2.5 — Two-way and intermediate control of a lighting point.

Materials

3/·029 V.R.I. Red.
3/·029 V.R.I. Black.
14/·0076 Flexible cord.
2 5A two-way switches.
1 5A intermediate switch.
1 Two-plate ceiling rose.
1 B.C. lamp holder.
4 Hard wood round blocks.

Diagrams

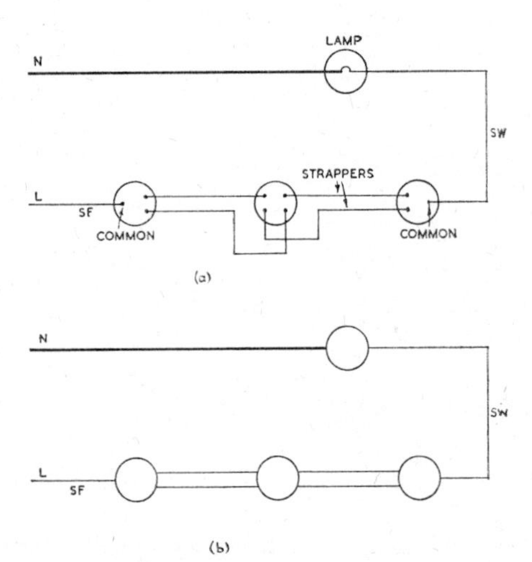

FIG. 2.11 (a–b) Exercise 2.5
(a) Connections; (b) Layout.

Procedure

Connect the circuit as shown in Fig. 2.11 (a) and (b).

Answer the following questions

1. Explain the meaning of the terms 'switch feed' and 'switch wire'.
2. Draw a neat diagram clearly showing the contact positions in the intermediate switch.

EXERCISE
No. 2.6

Master switch control.

Materials

3/·029 V.R.I. Red.
3/·029 V.R.I. Black.
14/·0076 Flexible cord.
2 5A one-way lighting switches.
2 5A two-way lighting switches.
1 Two-plate ceiling rose.
1 B.C. lamp holder.
5 Hard wood round blocks.

Diagrams

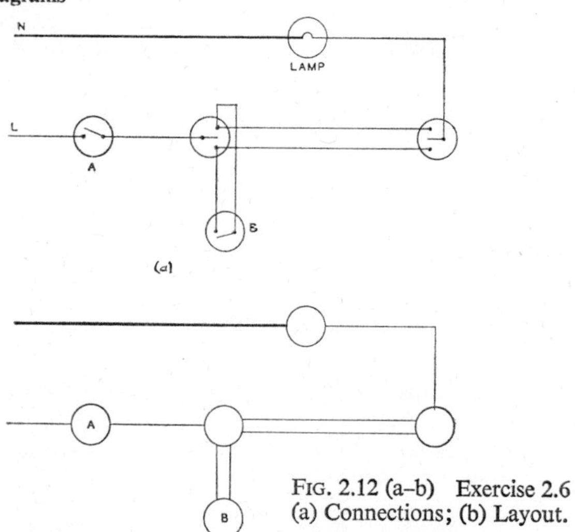

FIG. 2.12 (a–b) Exercise 2.6
(a) Connections; (b) Layout.

Procedure

Connect the circuit as shown in Fig. 2.12 (a) and (b).

Answer the following questions

1. Why does switch A prevent the light from being switched on by either of the two-way switches?
2. Why does switch B prevent the light from being switched off by either of the two-way switches?

CHAPTER 3

Wiring Systems - I

3.1 Types of Wiring System

(*a*) There are many systems of wiring which can be used to provide a safe, efficient and economical installation. Each of the recognized systems of wiring has its own particular merits, and the system used for any particular installation should be chosen with regard to the following factors:

(i) Type of load to be supplied.
(ii) Type of building in which the installation is situated.
(iii) Cost.
(iv) Durability.
(v) Appearance.
(vi) Any special adverse conditions (e.g. presence of inflammable vapours, risk of mechanical damage, etc.)
(vii) Expected life of installation.

(*b*) Some of the principal wiring systems in use at the present time are listed below:

(i) Bare conductor systems.
(ii) Cleated wiring.
(iii) All insulated wiring.
(iv) Lead alloy sheathed V.R.I. cables.
(v) Mineral insulated copper sheathed cables.
(vi) Earthed concentric wiring.
(vii) Catenary supported wiring.
(viii) Paper insulated cables.
(ix) Steel conduit and insulated conduit systems.
(x) Cable trunking and bus-bar trunking systems.
(xi) Underfloor and concealed duct systems.

3.2 Bare Conductors

(*a*) The I.E.E. Regulations state that bare or lightly insulated conductors may be installed in buildings for the following purposes only:

(i) Earthing connections.
(ii) The external conductors of earthed concentric systems.
(iii) The conductors of extra low voltage systems.
(iv) Protected rising main and bus-bar systems.
(v) Collector wires for travelling cranes, trolleys, etc.

(*b*) When installing bare conductors the following points must be observed if the system is to comply with the relevant I.E.E. Regulations and Codes of practice:

(i) In an extra low voltage system the insulation must be adequate, and protection against fire risk must be provided.

(ii) In rising main and bus-bar systems the conductors should be inaccessible to unauthorized persons. The insulators used to support the bus-bars must be strong enough to support them firmly, but at the same time the conductors must be free to expand and contract due to temperature changes.

(iii) Where bare conductors pass through floors, walls, partitions or ceilings they must be protected by enclosing them in non-absorbent incombustible insulating material, unless earthed metal trunking is used

(iv) Collector wires for travelling cranes, etc., should be protected by screens or barriers, unless there is adequate clearance to afford protection against inadvertent contact. Notices reading 'DANGER – BARE LIVE WIRES' in red letters on a white background should be fixed at each end of the collector wires, and throughout their length at intervals not exceeding 12 yards.

3.3 Cleated Wiring

(*a*) This wiring system is carried out using single-core insulated cables supported by insulating cleats. The system is relatively inexpensive but it affords little protection against mechanical damage and so it is useful only in dry situations where the risk of mechanical damage is small.

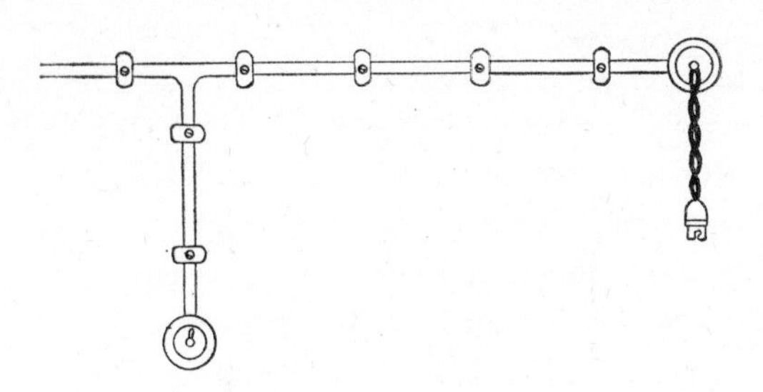

FIG. 3.1 Cleated Wiring

(*b*) When installing this type of wiring the following requirements of the I.E.E. Regulations should be observed:

(i) The cable should be supported on insulators having smooth or rounded edges which are spaced so as to prevent the cables coming into contact with each other or any other object.

(ii) The cables used should be rubber-insulated braided and compounded or P.V.C. insulated.

(iii) The cables should be open to view throughout their length except that where it is necessary to install cables under floors or within walls, partitions, etc., or where they are exposed and less than 6 ft. above floor level, additional mechanical protection is necessary. It is advisable to use non-metallic conduit for this protection because of the difficulty of providing efficient earthing for short lengths of metallic conduit in an otherwise all-insulated system.

(iv) The cables should pass directly through walls, floors, partitions, ceilings, etc., and should be protected by being enclosed in an incombustible conduit with suitably bushed ends.

(*c*) When installing cleated wiring the cable runs should be marked out and the cleats loosely fixed in position. The cables should be placed in the first cleat which is then firmly tightened; the cables are then slipped into the next cleat and strained while the cleat is tightened. This process is continued along the run so that the cables pass neatly through the cleats without sagging. Fittings such as switches and ceiling roses are usually mounted on hard wood blocks.

3.4 All-Insulated Wiring Systems

(*a*) A reasonably inexpensive method of wiring is to use cables which consist of one, two or three insulated cores enclosed in a sheath of tough insulating material which provides the necessary mechanical protection. All-insulated wiring systems are often used in the smaller type of installations for both lighting and heating circuits as this is probably the most economical system for this class of work. The all-insulated systems are not usually considered sufficiently robust for industrial installations where extra hazards such as high temperatures or increased risk of mechanical damage may exist. The principal types of cable in use at the present time are:

(i) *Tough rubber-sheathed* (*T.R.S.*) *cables*. Vulcanized rubber-insulated conductors which are protected by a tough rubber sheath.

(ii) *Polyvinyl-chloride-sheathed* (*P.V.C.*) *cables*. There are two types, P.V.C./P.V.C. in which P.V.C.-insulated conductors are used, and P.V.C. Polythene in which polythene insulated conductors are used. In both cases mechanical protection is provided by a tough P.V.C. sheath.

(iii) *Polychloroprene-sheathed* (*P.C.P.*) *cables*. These cables may be used where conditions are too arduous for P.V.C. or T.R.S. cables. Polychloroprene (or neoprene) is resistant to oil and petrol and can be used where there is exposure to steam, sulphur fumes, ammonia fumes, lactic acid, direct sunlight and heat. This type of cable is very suitable for farm installations, particularly in the dairy.

(iv) *House service overhead system cable* (*H.S.O.S.*). These cables are suitable for outside wiring, such as on external walls or supported by a catenary between buildings. They are resistant to the adverse effects of weather and exposure to direct sunlight.

(*b*) In order to comply with the I.E.E. Regulations the following points must be observed:

(i) Whenever possible the cores of the cable should be distinguished by colours in accordance with Table 7 of the I.E.E. Regulations. However, situations arise where it is not possible to adhere strictly to the recommendations of the table when using sheathed cables, a common case being when connection is made to a switch. The twin cable employed for this purpose will have its cores coloured red and black so that inevitably one 'live' core must be coloured black. In this situation the red core must be used for the switch feed leaving the black core for the switch wire. Wherever it has been necessary to indicate the colours of cables in the diagrams in this book the scheme shown in Fig. 3.2 is used.

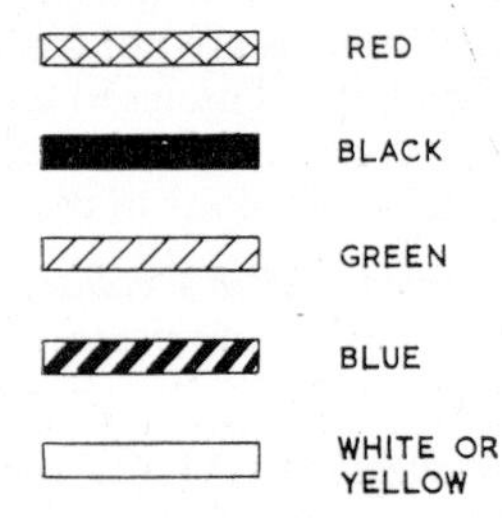

FIG. 3.2 Scheme for Indicating Colours of Cables

(ii) At switches, socket outlets, lighting fittings, and joint boxes the sheath of the cable must terminate inside a box, or in an incombustible enclosure formed by part of the accessory or fitting and the building structure.

(iii) When the cable is bent it must form a radius not less than four times the outer diameter of the cable.

(iv) When installed in accessible positions cables must be supported by clips not exceeding those set out in Table 9 of the I.E.E. Regulations. When installed in situations where they are unlikely to be disturbed, horizontal runs may rest without fixing provided the surface is dry and reasonably smooth. A vertical run without intermediate support must not exceed 15 ft. and the cable should be held at the top by means of a rounded support of radius not less than four times the outer diameter of the cable.

(v) When supported by a separate catenary wire the cable must be either continuously bound up with the catenary wire or attached to it at intervals not more than twice those set out in Table 9 of the I.E.E. Regulations.

(vi) Rubber P.V.C. or Polythene cables must not be installed in situations where the ambient temperature is likely to exceed 45° C.

(vii) Rubber-sheathed cables should not be installed in situations where they are exposed to direct sunlight unless they are protected by a semi-embedded braid applied during the process of manufacture or by a specially treated tape.

(viii) Where a cable passes through structural metal work, the hole must be bushed to prevent abrasion of the cable.

(*c*) When an installation is wired using multicore cables the necessary connections may be made using either joint boxes or three-plate ceiling roses.

(i) *Joint box method.* Fig. 3.3 (a) illustrates how a joint box may be used to make the connections required for a simple lighting circuit.

(ii) *Three-plate ceiling rose method.* The three-plate ceiling rose provides a good alternative to the use of joint boxes; the method is illustrated by Fig. 3.3 (b).

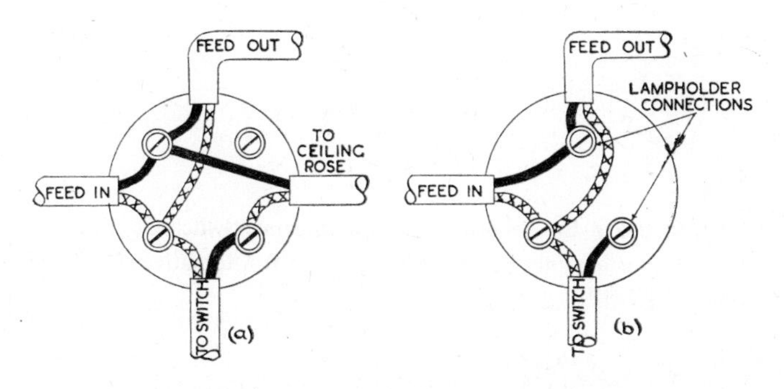

FIG. 3.3 (a–b) Joints in All-Insulated Cable

(iii) *Two-way and intermediate switching.* When using sheathed cables to wire a two-way and intermediate switched circuit a difficulty may arise when using the circuits given in Chapter 2. This is because it becomes necessary to provide some form of connector behind one switch block to make a joint in the switch line, Fig. 3.4 (a). This difficulty can be avoided by using the circuit shown in Fig. 3.4 (b).

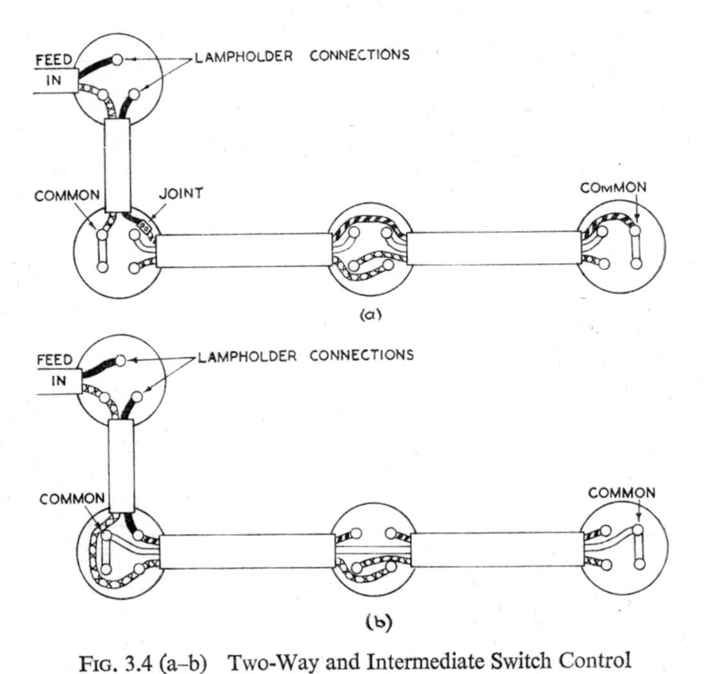

FIG. 3.4 (a–b) Two-Way and Intermediate Switch Control

(a) Circuit using joint behind switch block; (b) Alternative circuit which avoids extra joint.

3.5 Lead Alloy Sheathed V.R.I. Cable

(*a*) This type of cable has conductors insulated by vulcanized rubber, the mechanical protection being provided by a lead alloy sheath. This system provides an alternative to the use of all-insulated sheathed wiring (para. 3.4). The lead sheath provides somewhat greater protection against the ingress of moisture, nevertheless although widely used in the past the system is less often used in modern installations.

(*b*) Although in general the requirements of the I.E.E. Regulations for the lead-sheathed V.R.I. cable are the same as for all-insulated wiring systems the following differences should be noted:

(i) The minimum radius of a bend must not exceed six times the outer diameter of the cable.
(ii) The metal sheath must be effectively earthed and, if non-metallic joint boxes are used, means must be provided to maintain earth continuity.
(iii) The metal sheath must be prevented from coming into contact with any part of a wiring system operating at extra low voltage or any radio, telephone, bell or sound distribution system.
(iv) Where practical the metal sheath should be prevented from coming into contact with gas or water pipes or non-earthed metalwork. Alternatively, if this is not possible, the lead sheaths must be bonded to the other services or non-earthed metal so as to prevent any appreciable difference of potential at points of contact.

(*c*) Connections between cables can be made using either joint boxes or three-plate ceiling roses as in the all-insulated systems but, as it is necessary to maintain earth continuity throughout the system metal bonding clamps must be employed, as illustrated in Fig. 3.5.

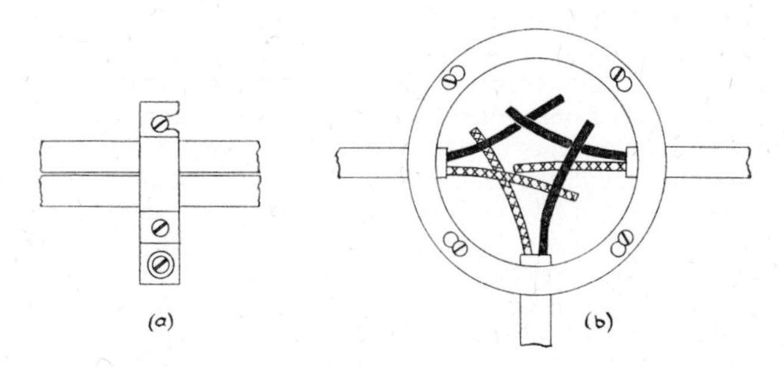

FIG. 3.5 (a–b) Bonding of Lead Alloy Sheathed Cable
(a) Bonding clamp; (b) Bonding ring.

3.6 Mineral Insulated Copper Sheathed Cables (M.I.C.S. Cable)
(*a*) This is a comparatively modern wiring system in which the copper

conductors are insulated by compressed magnesium oxide and enclosed in a seamless copper tube, Fig. 3.6.

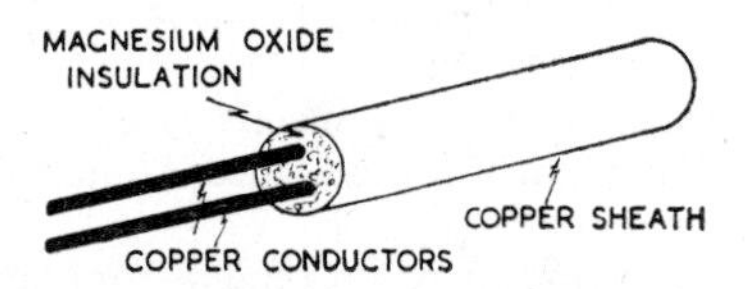

FIG. 3.6 M.I.C.S. Cable

Because of the mineral insulation the cable can withstand high temperatures, and the copper outer sheath is impervious to moisture, is unaffected by oil, and resists the action of most chemicals. Another advantage of this type of cable is its comparatively high current rating, which is due to the heat-resisting properties of the cable and to the fact that the magnesium oxide is a reasonably good heat conductor. This type of cable is not often used for domestic wiring because of its cost, but its many good properites can often be used to advantage in industrial installations, for example, in boiler houses and other hot situations where the heat-resisting properties are of particular value. M.I.C.S. cables are suitable for use in situations where flammable or explosive dust, vapour or gas is likely to be present; a further advantage in garages is that the cable is resistant to the effects of oil and petrol. As the cables are small in diameter, and can be readily bent to follow most contours, they can provide a very neat and unobtrusive installation, extra mechanical protection being needed only where exceptional risk of damage exists.

(*b*) The requirements of the I.E.E. Regulations for this type of cable are in general the same as for lead-sheathed cables with the following addition: The ends of the cable must be sealed against the ingress of moisture using a suitable insulating sealing compound, all moisture being expelled from the insulation before the sealing compound is applied. The reason for this is because the magnesium oxide insulation is hygroscopic, i.e. it tends to attract and absorb moisture if left exposed to the atmosphere, and this absorbed moisture tends to lower the insulation resistance.

(*c*) A typical method of sealing the cable is by fitting a 'Pot' type seal as described below:

(i) Cut through the copper sheath at the required termination position with a 'ringing tool'. (Fig. 3.7.)

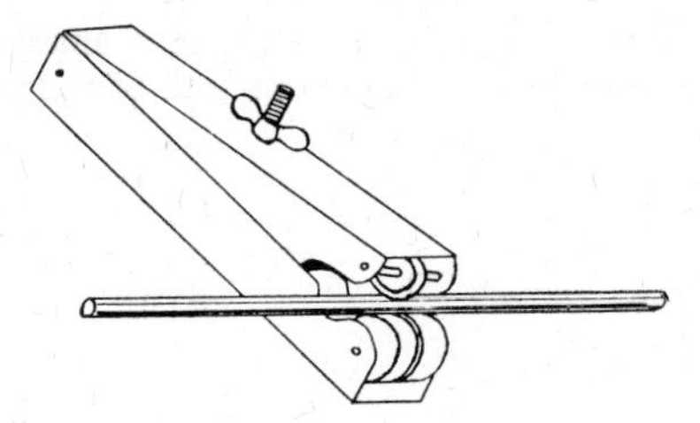

FIG. 3.7 Ringing Tool

(ii) Remove the unwanted copper sheath from the end of the cable using either side cutters or a special stripping tool. (Fig. 3.8.)

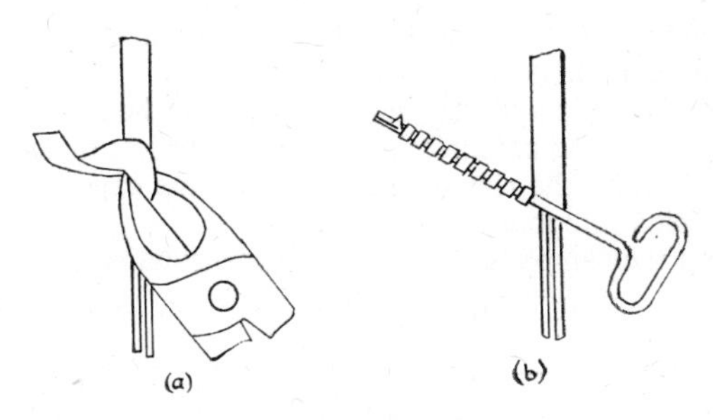

FIG. 3.8 Stripping M.I.C.S. Cable

(a) Using side cutters; (b) Using special tool.

(iii) Ensure that all insulation is removed from the conductors right back to the end of the sheath and that any burrs are removed.

(iv) Pass the appropriate gland over the cable end and screw on the sealing pot until the end of the sheath is level with the pot. Ensure that any dirt or metal is removed from the pot.

(v) Pass suitable lengths of sleeving through the holes in the sealing disc (note that sleeving can be obtained with moulded anchoring collars at one end (Fig. 3.9 (a)), alternatively an anchoring wedge in the form of a small bead may be inserted to make a bulge in one end of the sleeving). (Fig. 3.9 (b)).

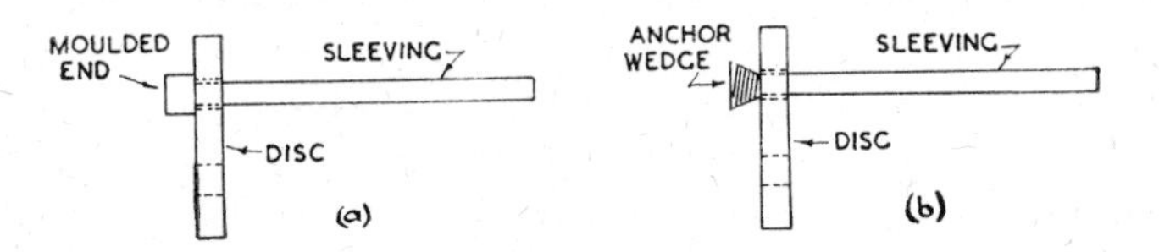

FIG. 3.9 (a–b) Sealing Disc and Sleeving Assembly
(a) Sleeving with moulded anchoring collar; (b) Use of anchoring wedge.

(vi) Slide the cap and sleeving assembly over the conductors to make sure that the conductor spacing is correct. Withdraw the sleeving assembly just enough to enable the sealing compound to be pressed into the sealing pot, and fill the pot with compound.

(vii) Press the cap and sleeving assembly into the pot and use the crimping and compression tool, to compress the compound and crimp the rim of the pot on to the cap. Remove the surplus compound.

(viii) Test the completed termination for insulation resistance.

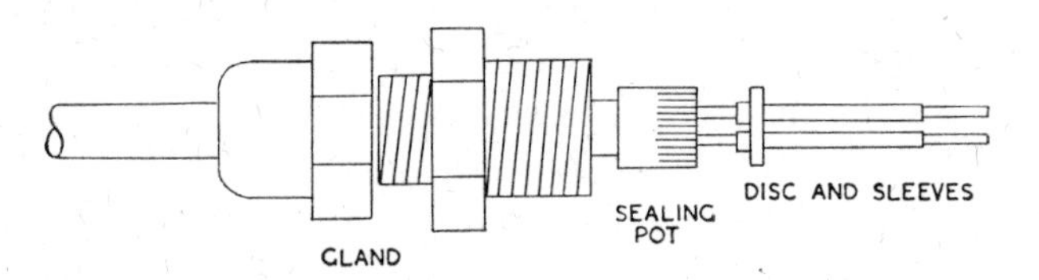

FIG. 3.10 Termination of M.I.C.S. Cable

3.7 Earthed Concentric Wiring

(*a*) This system, which is seldom used nowadays, consists of a single insulated conductor protected by an earthed metal sheath which is also used as the neutral or return conductor.

(*b*) This system can be used only where it is possible to comply with the following I.E.E. Regulations:

(i) It is supplied by a transformer, convertor or private generating plant in such a manner that there is no metallic connection with the public supply, or it is connected to an a.c. supply system on which multiple earthing has been authorized.

(ii) The external conductor shall be earthed.

(iii) No fuse, or non-linked switch or circuit breaker shall be inserted in the earthed external conductor.

3.8 Catenary Supported Wiring

(*a*) Cables can be supported between buildings by means of a high tensile galvanized steel catenary wire, this provides a useful method of extending a supply from a building to an outhouse. For example, from a domestic dwelling to a garden shed or garage.

(*b*) In order to comply with the I.E.E. Regulations cables used for this purpose must be suitable for outdoor use such as H.S.O.S., P.C.P. or T.R.S. with partially embedded braid and overall compound.

The weight of the conductors must be fully supported by the catenary wire. The I.E.E. Regulations recommend that cables supported by a separate catenary wire shall either be continuously bound up with the catenary wire or attached to it at intervals not more than twice those set out in Table 9 in the case of T.R.S. or P.V.C. sheathed cables. If lead-sheathed cables are so supported they should be attached at intervals not exceeding those set out in Table 9 of the I.E.E. Regulations.

(*c*) The following important points should be kept in mind when installing catenary supported cables:

(i) The catenary wire must be securely fixed at each end.

(ii) The clearance between the cables and the ground must be adequate; a suitable pole may be useful in some circumstances.

(iii) The difference in the levels of the ends of the catenary should not be excessive.

(iv) The cable, where it leaves the catenary, should pass through a suitable glazed porcelain lead-in tube.

(v) 'Drip loops' should be provided at entry points so that water is not led into the building by running along the cable.

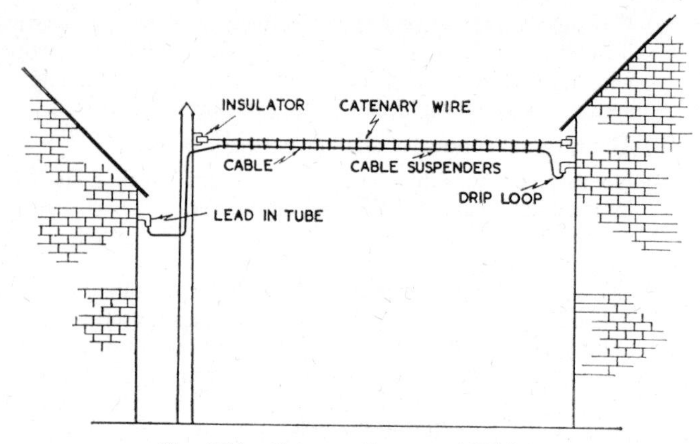

FIG. 3.11 Catenary Supported Wiring

3.9 Grid Suspension Wiring System

(*a*) A catenary system of wiring may be used in larger industrial buildings and similar situations where other systems would be difficult and expensive to install. The cables used have a number of insulated conductors combined with a catenary wire, a cross-section of a typical cable of this type is illustrated in Fig. 3.12.

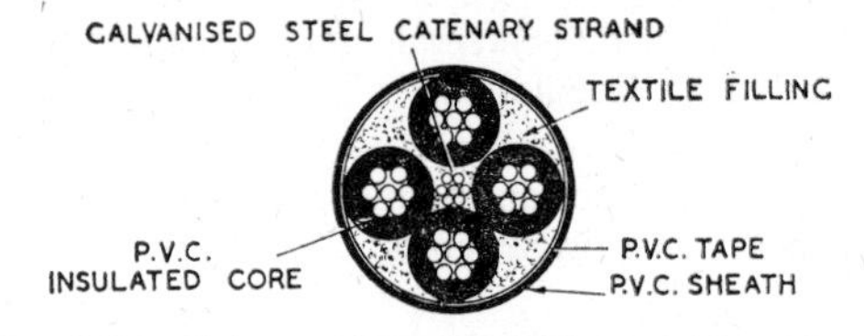

FIG. 3.12 Cross-section of Grid Suspension Cable

(*b*) When installing this type of system it is obviously important that the weight of the cables should be supported by the catenary wire so that there is little stress in the conductors. To this end the catenary wire is secured at each end by means of eye bolts and strainers. Special connecting boxes, incorporating catenary clamps, are used at tees, right angled turns, etc., and to support fittings. The boxes have fixing bridles which are used to support them from girders, ceilings, walls, etc. The types of cable normally employed are suitable for spanning up to about 170 ft., but it is usual to provide intermediate support from roof trusses, particularly where lighting fittings are suspended from the system. If the system is installed out of doors it is advisable to fill the connecting boxes with plastic compound to provide protection against the ingress of moisture. One standard type of connection box can be used for most purposes, some typical uses being shown by Fig. 3.13 (a–c).

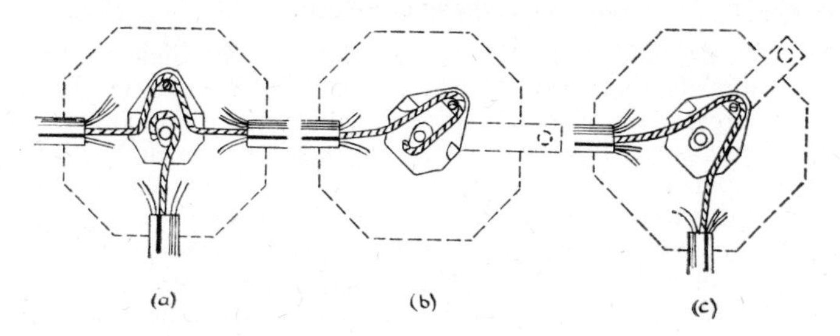

FIG. 3.13 (a–c) Use of Standard Connection Boxes
(a) Tee-off; (b) Span termination; (c) Right angle.

3.10 Paper Insulated Lead-Sheathed Cables

(*a*) This type of cable consists of one or more stranded copper conductors insulated by several layers of paper tape impregnated with insulating oil, the insulated conductors being enclosed in a lead sheath which is impervious to moisture. Mechanical protection is afforded to the lead sheath in one of the following ways:

(i) A hessian serving, impregnated with a bitumastic compound.
(ii) Steel wire armouring.
(iii) Steel tape armouring.

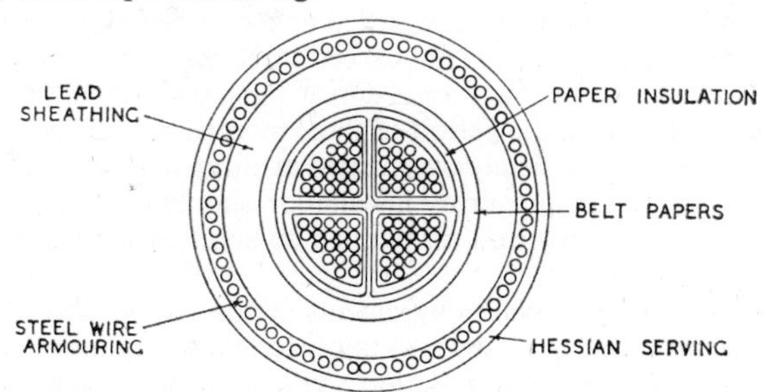

FIG. 3.14 Paper Insulated Lead-Sheathed Cable

This type of cable is mainly used for underground supply mains, but it can be used with advantage in situations where a large current carrying capacity is required, or for extensions to an installation which involve running the cables underground, e.g. for wiring between two buildings.

(*b*) The main requirements of the I.E.E. Regulations particularly applying to the use of this type of cable are:

(i) Terminations and joints should be made using correctly designed joint boxes, these can be filled with compound to prevent the ingress of moisture. In practice the compound not only helps to seal the cable against moisture but also holds the conductors firmly in position in the joint box and adds to the insulation.

(ii) The radius of any bends made in the cable should not be less than 12 times the overall diameter of the cable.

(iii) Where the cable is so installed that drainage of the impregnated compound is liable to occur, e.g. for rising main installations, cables complying with the drainage test of B.S.480 must be used.

(iv) When cables are installed underground they should be installed at a depth of at least 18 in.

3.11 Temporary Installations

(*a*) A temporary installation may sometimes be required to provide lighting or power for a limited period, for example, during the construction of a building.

(*b*) The requirements of the I.E.E. Regulations for such an installation are:

(i) An installation can be regarded as temporary if its expected period of service does not exceed three months.

(ii) When it is necessary to retain a temporary installation in service beyond the period of three months initially estimated it shall be completely overhauled at three-monthly intervals.

Note 1. – A temporary installation should be disconnected from the supply and dismantled as soon as it is no longer required.

Note 2. – Special requirements of the local authority and the insurance company may apply to temporary installations.

(iii) Every temporary installation shall be adequately protected against excess current and shall be effectively controlled by a conveniently situated switch or other means, whereby all phases or poles of the supply, including the neutral, can be disconnected when the installation is not in use. Where a temporary installation is to be supplied from a permanent installation, the current rating of the permanent installation must be adequate for the load to be imposed upon it.

(iv) In any temporary installation the total load on a final sub-circuit to which bayonet-socket lamp holders are connected shall not exceed 1,000 watts.

Note. – All-insulated lamp holders (which should preferably be fitted with a skirt) should be used.

(v) Except in a private dwelling-house, every temporary installation shall be in the charge of a competent person, who shall accept full responsibility for the installation, for its use, and for any alteration or extension. The name and designation of this person shall be prominently displayed close to the main switch or circuit-breaker.

(vi) All cables in a temporary installation shall normally be sheathed with tough-rubber or P.V.C. or be steel-armoured, but in positions where the cable is not liable to be handled or exposed to mechanical damage, rubber-insulated and braided and compounded or P.V.C.-insulated cable may be used.

(vii) Metal-sheathed cables shall not be used unless they are armoured. If steel conduit is used, the installation shall conform with Regulations 218 and 219. The insulation of the cables shall be in good condition.

(viii) Flexible cords in a temporary installation shall comply with the relevant Regulations and shall be used only where essential. Wherever exposed to the risk of mechanical damage they shall be tough-rubber or P.V.C.-sheathed.

(ix) A temporary installation shall be tested before it is put into service, and shall comply as regards insulation resistance, correctness of polarity and earth-continuity with the requirements of Section 5 of the I.E.E. Regulations.

(*c*) When planning a temporary installation, cost and speed of erection are important factors, governing the choice of system to be used. T.R.S. or P.V.C. cables are usually employed, often being fixed by suitable binding to convenient parts of the structure. It cannot be too strongly emphasized that the hazards of this type of installation may be far greater than in a normal installation, for example, cables may be liable to mechanical damage, due to the nature of the work being carried out on the site. In many cases the use of armoured cables must be considered and extra protection may be needed, for example, by using suitably placed wooden duck-boards. Shock risks are often high because of the presence of wet conditions and of metal scaffolding, so great care must be taken with earthing arrangements and the siting of lighting fittings, switches, etc. Watertight connections for trailing leads are often desirable. In particularly dangerous situations the use of a low voltage system should be considered. A 'competent person' in charge of a temporary installation has a very high degree of responsibility; especially if an accident should occur, and so should make certain that the installation is always maintained in a satisfactory and safe condition.

EXERCISES

The exercises which conclude this chapter are designed to illustrate some of the commonly used wiring systems.

EXERCISE
No. 3.1

Two lighting points each separately controlled, wired in all-insulated cable using joint box method.

Materials

3/·029 Twin T.R.S. or P.V.C. cable.
14/·0076 Flexible cord.
1 Bakelite joint box.
2 Two-plate ceiling roses.
2 B.C. lamp holders.
2 5A one-way lighting switches.
4 Hard wood round blocks.

Layout

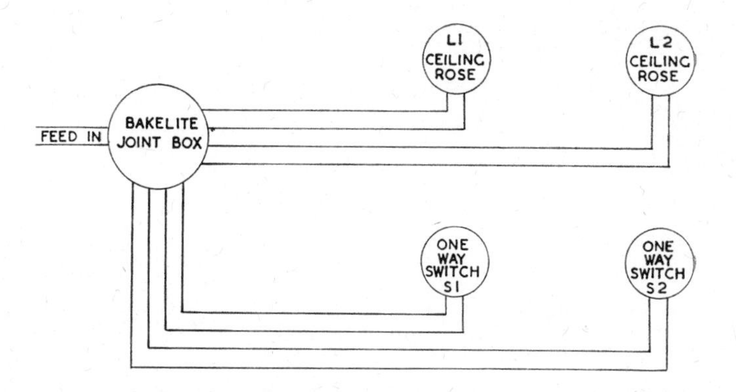

FIG. 3.15 Exercise 3.1

Procedure

Connect the circuit so that S1 controls L1 and S2 controls L2. (Fig 3.15.) Draw a diagram showing clearly the connections inside the joint box.

Answer the following questions

What are the requirements of the I.E.E. Regulations regarding:

(i) The colours of the cores of the cables used in this exercise?
(ii) The minimum radius of a bend in T.R.S. cable?

EXERCISE No. 3.2 **Three lighting points each separately controlled, wired in all-insulated cable, using three-plate ceiling roses.**

Materials

3/·029 Twin T.R.S. or P.V.C. cable.
14/·0076 Flexible cord.
3 Three-plate ceiling roses.
3 B.C. lamp holders.
3 5A one-way lighting switches.
6 Hard wood round blocks.

Layout

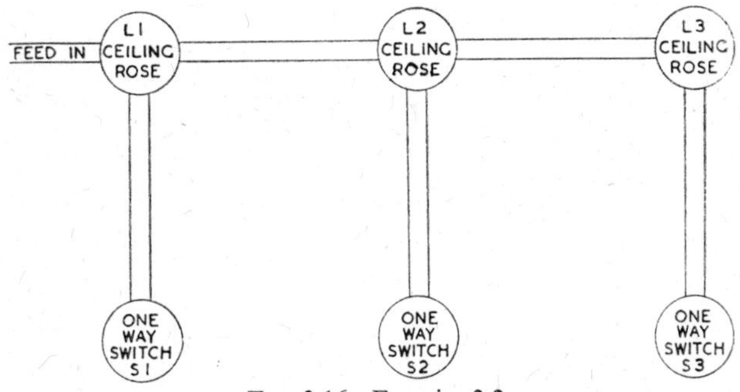

FIG. 3.16 Exercise 3.2

Procedure

Connect the circuit so that S1 controls L1 S2 controls L2 and S3 controls L3. (Fig. 3.16.)

Draw a diagram showing clearly the connections at ceiling rose No. 1.

Answer the following questions

What are the requirements of the I.E.E. Regulations regarding:

(i) Enclosure of the cores of sheathed cables from which the sheath has been removed?
(ii) Exposure of T.R.S. cable to sunlight?

EXERCISE No. 3.3

Lighting point controlled by 2 two-way and one intermediate switch wired in all-insulated cable, using a three-plate ceiling rose.

Materials

3/·029 Twin T.R.S. or P.V.C. cable.
3/·029 Three-core T.R.S. or P.V.C. cable.
14/·0076 Flexible cord.
1 Three-plate ceiling rose.
1 B.C. lamp holder.
2 5A two-way lighting switches.
1 5A intermediate lighting switch.
4 Hard wood round blocks.

Layout

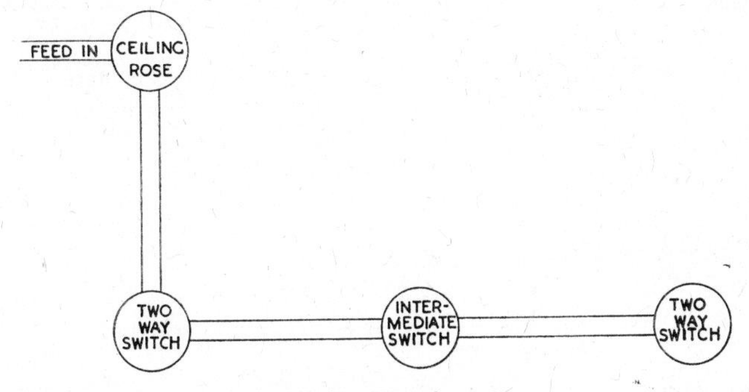

FIG. 3.17 Exercise 3.3

Procedure

Connect the circuit to provide two-way and intermediate switch control of the lighting point. (Fig. 3.17.) Draw a diagram showing clearly the connections made to the two-way and intermediate switches.

Answer the following questions

What are the requirements of the I.E.E. Regulations with regard to:

(i) P.V.C. or T.R.S. cables installed in very hot situations?
(ii) Holes where the cable passes through structural metalwork?

EXERCISE NO. 3.4

Lighting point controlled by two two-way switches wired in lead-sheathed V.R.I. cable using the joint box method.

Materials

3/·029 Twin lead-sheathed V.R.I. cable.
3/·029 Three-core lead-sheathed V.R.I. cable.
14/·0076 Flexible cord.
1 Metal joint box.
1 Bonding ring.
1 Two-plate ceiling rose.
1 B.C. lamp holder.
2 5A two-way lighting switches.
3 Hard wood round blocks.

Layout

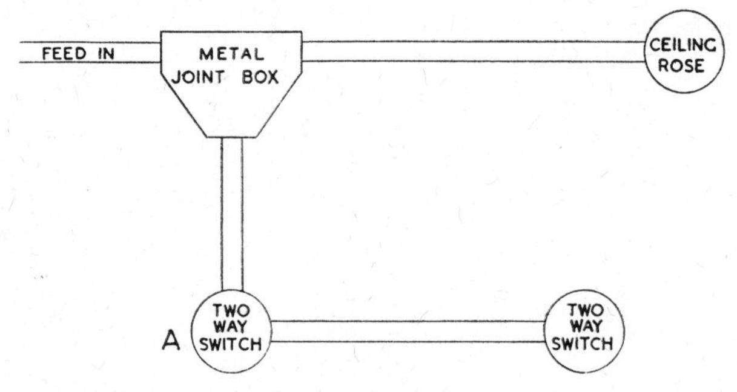

FIG. 3.18 Exercise 3.4

Procedure

Connect the circuit to provide two-way switch control of the lighting point. (Fig. 3.18.)

Draw diagrams showing clearly how effective earth continuity is maintained at (*a*) the joint box and (*b*) the two-way switch at A.

Answer the following questions

What are the requirements of the I.E.E. Regulations with regard to:

(i) Earthing of the metal sheath?
(ii) Contact of the lead sheath with other services?

Exercise No. 3.5 — Lighting point controlled by 2 two-way switches using M.I.C.S. cable.

Materials

0·0015 in^2. Twin M.I.C.S. cable.
0·0015 in^2. Three-core M.I.C.S. cable.
14/·0076 Flexible cord.
3 Sealing pots and glands for 0·0015 in^2. twin M.I.C.S. cable.
2 Sealing pots and glands for 0·0015 in^2. three-core M.I.C.S. cable.
1 $\frac{3}{4}$-in. E.T. conduit through box.
1 $\frac{3}{4}$-in. E.T. conduit angle box.
1 $\frac{3}{4}$-in. E.T. conduit terminal box.
1 Ceiling rose.
2 5A two-way lighting switches.
1 B.C. lamp holder.

Layout

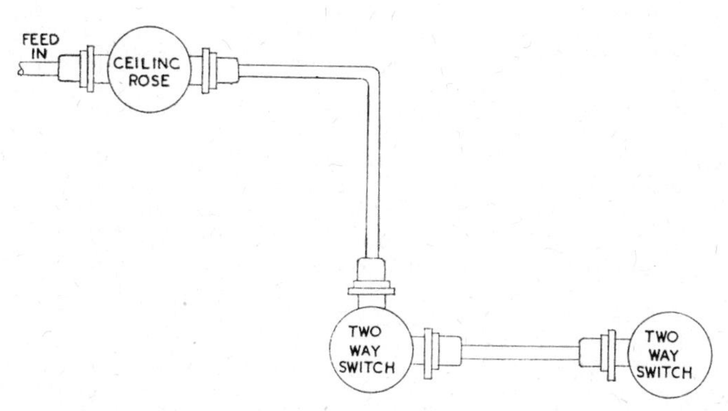

FIG. 3.19 Exercise 3.5

Procedure

Connect the circuit to provide two-way switch control of the lighting point. (Fig. 3.19.)

Draw a sketch of a complete M.I.C.S. cable termination.

Answer the following questions

1. Why is the current rating of an M.I.C.S. cable higher than that of a rubber- or P.V.C.-insulated cable of the same cross-sectional area?
2. Give three examples of situations where M.I.C.S. cable can be used to advantage.

EXERCISE No. 3.6

Termination of paper insulated lead-covered double steel taped armoured cable (P.I.L.C.D.S.T.A.).

Materials

0·0225 in^2. Two-core P.I.L.C.D.S.T.A. cable.
1 Termination box.
2 0·0225 in^2. weak back jointing sleeves.
Black 7/·064 V.R.I. taped and braided cable.
Red 7/·064 V.R.I. taped and braided cable.
Demonstration board No. 1.

Procedure

Carry out the steps illustrated by the demonstration board as explained below:

1. Fix the back of the termination box to the wall and lay the cable through the centre line with the end of the cable level with the top of the box. Mark off the serving 6 in. below the box. Place a wire binder around the cable at this point, and remove the serving above the binder.
2. Cut through the armouring at a point 1 in. below the bottom of the box and remove it. Open out the armour to the binder.
3. Cut away the jute bedding and thoroughly clean the exposed lead sheath.
4. Place the cable vertically alongside the box and mark off the lead sheath at the point where it will enter the box. Carefully cut, and remove with the aid of a hack knife, the lead sheath from the end of the cable to within $\frac{1}{4}$ in. of the mark already made.
5. Bind the belt paper immediately above the lead sheath with a layer of oil impregnated tape, and remove the belt papers above the tape.
6. Splay out the cores and cut out the loose jute wormings. Protect the cable cores by taping in the same direction as the core papers for a distance of approximately 1 in., using oil impregnated tape. Cut the cores off $\frac{3}{4}$ in. above the tape, and remove paper insulation above the tape.
7. Slip the cast-iron gland over the cable end.
8. Prepare the ends of the V.R.I. tails and tin the ends of the cable cores. Fit the weak back jointing sleeves over the cable cores and V.R.I. ends, and crimp them together, making sure that the ends of the conductors are well butted. Sweat the joints, and while the solder is still hot crimp the sleeves tightly on to the conductors.
9. Lay the cable in position in the box and bolt on the front half of the box.
10. Place the lead cone grip in position, and bolt up the gland.

11. Clean the armour and cast-iron gland thoroughly. Place the armour in position round the gland and clamp it tightly with the armour clamp.
12. Drill the wood bush to take the V.R.I. tails.
13. Fill the box with bitumen and fit the wood bush in position.

Answer the following questions

1. Give two examples where this type of cable is likely to be used.
2. What are the requirements of the I.E.E. Regulations concerning the termination and jointing of this type of cable?

Wiring Systems - II

4.1 Metal Conduit Systems

(*a*) A common type of installation is one in which the circuits are wired using single core insulated cables, these cables being afforded mechanical protection by enclosing them in metal conduit tubes. The advantages of this type of system are that it gives the cables a high degree of mechanical protection, minimizes fire risks, provides efficient earth continuity and allows easy rewiring. Short lengths of conduit can be used to provide additional mechanical protection where necessary to all-insulated and similar systems.

Types of metal conduit in common use are:

(i) Light gauge steel conduit.
(ii) Heavy gauge steel conduit.
(iii) Aluminium conduit.
(iv) Copper conduit.
(v) Flexible steel conduit.

(*b*) The principal requirements of the I.E.E. Regulations, and Code of Practice C.P.321:101, for the installation of conduits are:

(i) The conduit system must be completely erected and securely fixed before any cable is drawn in.
(ii) Inspection boxes, draw in boxes, etc., should be so situated that they remain accessible throughout the life of the installation.
(iii) The maximum number of cables must not be greater than the number given in Table 10 or 11 of the I.E.E. Regulations. For groups of cables not listed in the tables a space factor of 40% must not be exceeded.

$$\text{Space factor} = \frac{\text{Total cross-sectional area of installed cables}}{\text{Internal cross-sectional area of the conduit}} \times 100$$

If the cables are to be drawn round more than two 90° bends an appropriate reduction must be made in the maximum number of cables in order to facilitate the drawing in of the cables.

(iv) The inner radius of a conduit bend must not be less than $2\frac{1}{2}$ times the outside diameter of the conduit and should also be not less than 4 times the diameter of the largest cable installed.

(v) Solid elbows and tees may only be used at the ends of conduits immediately behind accessories or lighting fittings.

(vi) All burrs must be removed from the ends of lengths of conduit, and outlets must be bushed to guard against the possibility of abrasion of the cables.

(vii) Where joints in the cables are required substantial metal boxes complying with the appropriate British Standard (B.S.S.31) should be used.

(viii) At every termination the cables must be enclosed in an enclosure of incombustible material.

(ix) It is essential that the whole of the conduit system be efficiently earthed.

(x) As in the case of metal-sheathed cables metal conduit should not be allowed to come into contact with the metal work of other services, for example, gas or water pipes, or non-earthed metal work. If contact cannot be avoided all the metal work should be effectively bonded together.

(xi) When an a.c. supply is used the phase and neutral conductors of a circuit must be drawn into the same conduit.

(xii) Conduit systems not intended to be gas-tight shall be self-ventilating and drainage outlets should be provided wherever condensed moisture might collect.

(xiii) The cables of extra low voltage systems must not be drawn into the same conduits as the cables for power and lighting operating at a voltage exceeding extra low unless the cables used for the low voltage system are themselves insulated to withstand the highest voltage present in the conduit.

(xiv) Where the conduits themselves are particularly liable to mechanical damage adequate extra protection against damage should be provided.

(xv) Where conduits pass through floors, walls, partitions or ceilings the holes should be made good with cement or similar material. In addition, in circumstances where conduit passes through a danger area where flammable or explosive dust vapour or gas may be present, than a flame proof box must be inserted where the conduit enters a safe area.

(xvi) Where conduit is installed in a damp situation or is exposed to the weather, heavy gauge conduit should be used and this should have a corrosion resisting finish.

(xvii) Flexible metal conduit is not suitable for use as an earth continuity conductor, a separate earth continuity conductor being required.

4.2 Light Gauge Steel Conduit

(*a*) Light gauge steel conduit is manufactured to the requirements of British Standard Specification No. 31, 1940, Class A. Three types are available, close joint, brazed or welded joint and seamless or solid drawn. Standard sizes are: $\frac{1}{2}$ in., $\frac{5}{8}$ in., $\frac{3}{4}$ in., 1 in., $1\frac{1}{4}$ in., $1\frac{1}{2}$ in. and 2 in. outside diameter. As this type of conduit is not heavy enough to withstand screwing, there is difficulty in maintaining earth continuity through the fittings, particularly when using the older types such as 'slip-on', 'pin-grip' and similar fittings. Nowadays 'lug-grip' fittings, as illustrated in Fig. 4.1, are used, when using these fittings care must be taken to remove the paint or enamel from the end of the conduit to ensure that earth continuity is maintained. Although it is the least expensive of the conduit systems and is easy and quick to install, light gauge conduit is not suitable for use in damp situations and does not provide a very high degree of mechanical protection.

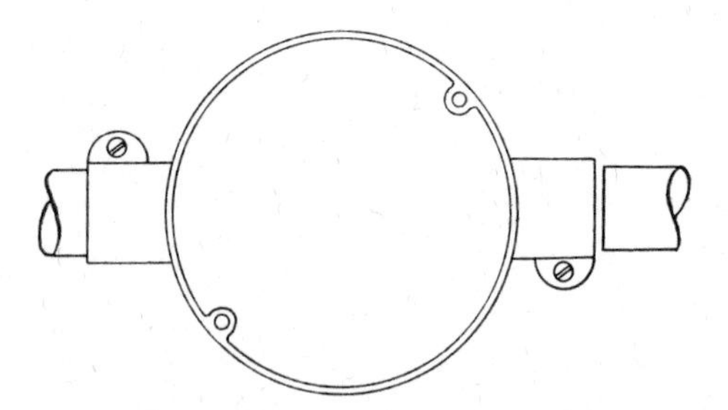

FIG. 4.1 Lug Grip through box

(*b*) Close joint conduit cannot be bent or set as the seam tends to open. Slight sets or bends may be made in brazed joint or solid drawn light gauge conduit if great care is used. Light gauge conduit can be cut using a hacksaw provided a fine-toothed blade is used. A good alternative is to score a deep groove all round the tube using a sharp file and then bend around the knee. After cutting the conduit the end must be filed to remove any sharp edges or burrs.

4.3 Heavy Gauge Steel Conduit

(*a*) Heavy gauge steel conduit is manufactured to the requirements of B.S.S.31, 1940, Class B. Two types are available, welded joint and solid drawn, the standard sizes being ½ in., ⅝ in., ¾ in., 1 in., 1¼ in., 1½ in., 2 in. and 2½ in. outside diameter. This class of conduit provides a robust installation and there is little difficulty in maintaining good earth continuity throughout the system as all joints to fittings are made by means of screwed threads. (Fig. 4.2.)

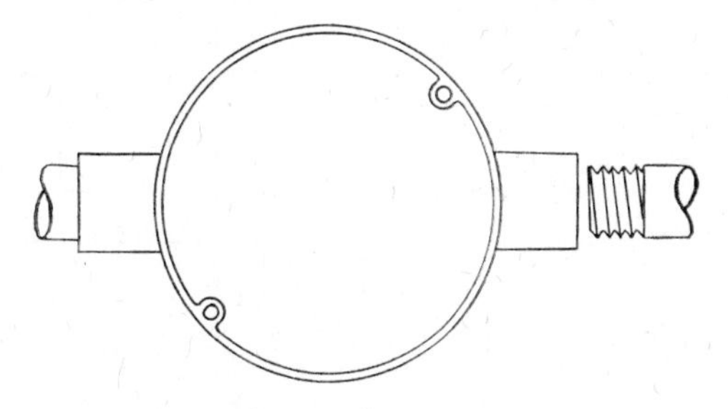

FIG. 4.2 Screwed Through Box

Some of the more common fittings available are shown in Figs. 4.3 (a-j).

(*b*) (i) Conduit is normally supplied in lengths of approximately 13 ft. When selecting a length of conduit for use, it is good practice to ensure that the bore is clear from any obstructions as these could lead to unnecessary waste of time and effort at a later stage. Sets and bends may be made using either a bending block or a bending machine. When using a bending block (Fig. 4.4 (a)), the conduit is inserted in the hole and a slight pressure exerted to bend it. The bend is made a little at a time, inching the conduit gently through the block until the desired degree of bend has been attained. Great

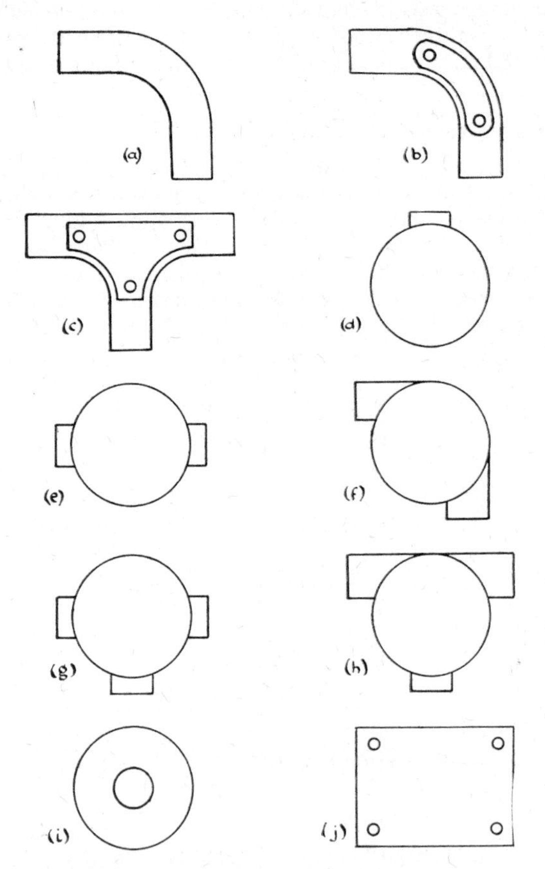

FIG. 4.3 (a–j) Screwed Conduit Fittings

(a) Solid Elbow; (b) Inspection Elbow; (c) Inspection Tee; (d) Terminal Box; (e) Through Box; (f) Angle Box; (g) Tee Box; (h) Tangent Tee Box; (i) Back Entry Box; (j) Adaptable Box.

care must be taken to avoid using excessive pressure and making too sharp a bend so flattening the conduit. When using a bending machine (Fig. 4.4 (b)) the conduit is placed in a grooved former, F, and held by a stop, S. When the lever is pulled the roller, R, presses on the guide, G, so forcing the conduit to take the shape of the former. When making a bend care should be taken not to bend through an excessive angle in the first instance, as it is easier to increase the amount of bend than to reduce it.

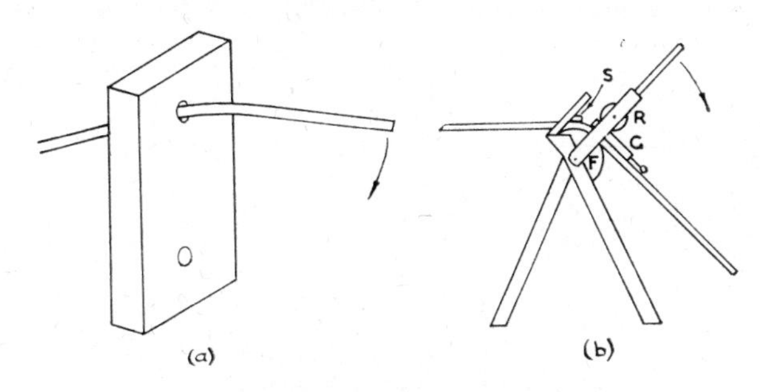

FIG. 4.4 (a–b) Bending Conduit
(a) Bending block; (b) Bending machine.

(ii) When cutting or threading conduit it is best held in a 'pipe' vice (Fig. 4.5) as using an ordinary vice with plain jaws tends to crush the conduit.

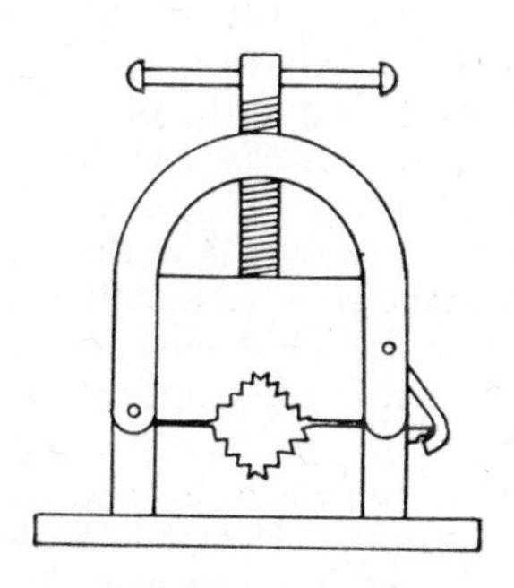

FIG. 4.5 Pipe Vice

A fine-toothed hacksaw is used for cutting, and before beginning to cut the thread all external burrs must be removed and the end of the conduit trued using a flat file. The end of the conduit is then lightly smeared with lubricant and the thread cut using conduit stocks and dies. The length of thread cut should be just sufficient to screw fully into the fitting to be used so that no thread will show after screwing up tightly. Finally, any burrs must be removed from inside the conduit using a reamer or round file, and any traces of lubricant or swarf wiped off (Fig. 4.6 (a–b)).

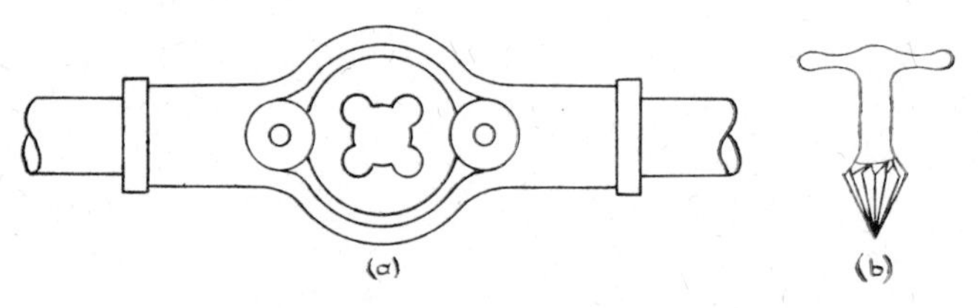

FIG. 4.6 (a–b) Conduit Tools
(a) Stock and die; (b) Reamer.

(*c*) When erecting a screwed conduit system the order of erection should be such that each succeeding length of conduit can be rotated so as to screw firmly into the preceding fitting or coupler. If it is not possible to rotate either of two adjacent lengths of tubing then a 'running coupler' can be used, the method of making this type of joint is explained in Exercise 4.1, at the end of this chapter. The number of running couplers used should be kept to a minimum, by careful planning of the order of erection of the conduit. Conduits run on the surface can be fixed using various types of saddle (Fig. 4.7 (a–d)). Ordinary saddles hold the conduit firmly against the wall, spacer bar saddles hold the conduit slightly away from the wall (by approximately $\frac{1}{8}$ in.) and their use helps to prevent corrosion due to damp walls and also avoids the need for frequent setting of the conduit where it enters fittings or accessories. A further advantage of spacer bar saddles is that only one fixing hole is required; and the saddle can be adjusted slightly up or down if the hole is a little out of alignment. Hospital saddles are used to give a greater spacing between conduit and wall, so helping to prevent the accumulation of dust and dirt behind the conduit. Girder clips are used where conduits must be fixed to girders or steel beams which cannot be drilled. Concealed conduit is often best fixed by using crampets.

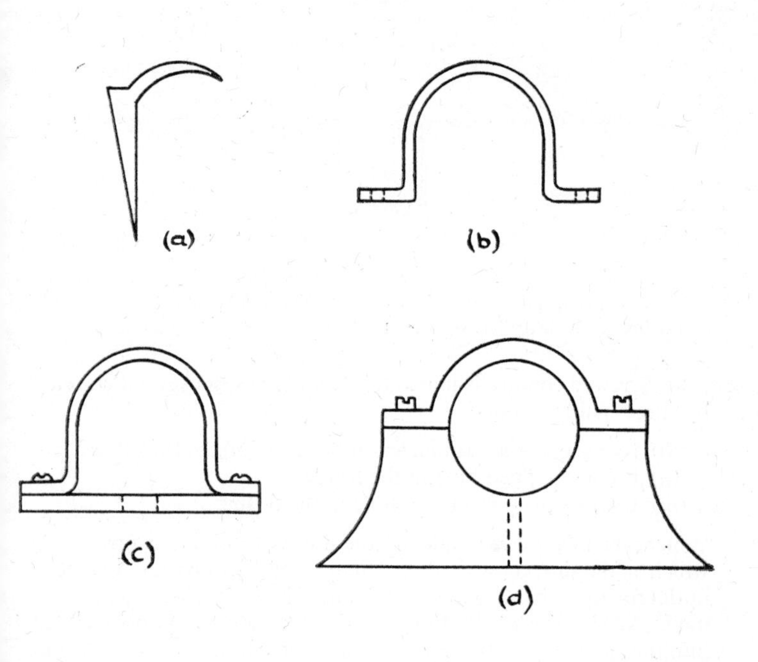

FIG. 4.7 (a–d) Conduit Fixings
(a) Crampet; (b) Saddle; (c) Spacer bar saddle; (d) Hospital saddle.

When erecting a conduit system great care should be taken to avoid moisture traps. Fig. 4.8 shows an example where the danger of a possible moisture trap can be avoided by using a conduit box with a $\frac{1}{8}$-in. hole drilled in the lid to allow moisture to escape.

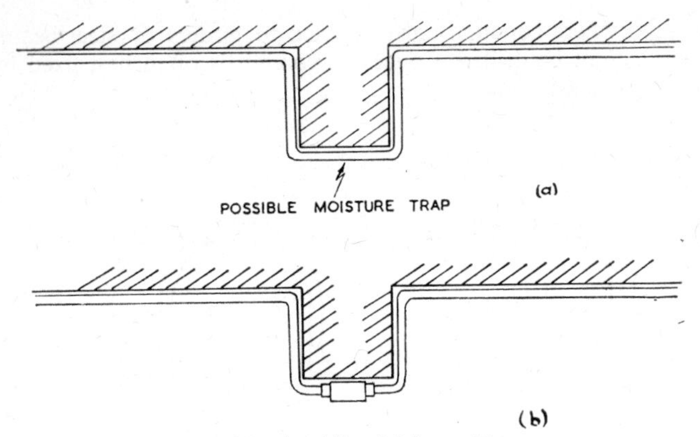

FIG. 4.8 Avoiding Moisture Traps

(a) Moisture trap caused by bends in conduit; (b) Moisture trap avoided by means of a hole drilled in a conduit box.

(*d*) Screwed conduit is joined to switch boxes, socket outlets, etc., by three main methods:

(i) By means of a screwed spout projecting from the fitting.
(ii) Using a tapped hole in the fitting.
(iii) Using a plain clearance hole in the fitting.

Whichever method is employed the aim is to obtain a firm fixing which maintains good earth continuity; and to provide a smooth outlet for the cables, so minimizing the risk of abrasion when drawing in. When a screwed spout is used, the conduit is simply screwed into the spout. With the other types of fixing either a male bush and coupler or a female bush and lock-nut may be employed, the latter method being preferred for fixing to clearance holes particularly with pressed steel fittings as a firmer fixing can be achieved.

(*e*) Before starting to pull in the cables it is advisable to ensure that the conduit system is free from obstructions and moisture. It is good practice to pull a swab through the conduit using a draw-wire, so removing any moisture and at the same time proving that there are no obstructions. In large conduit installations, it is better to commence pulling in the cables from a point near to the centre of a run, as this reduces the length of cable to be pulled in to a minimum. First, a steel tape (wireman's snake) can be passed through the conduit from one pull-in point to another. This is used to pull in a draw-wire which is subsequently used to pull in the cables, note that the 'wireman's snake' should not itself be used as a draw-wire, as the tape is easily damaged if subjected to excessive strains. When drawing in the cables, great care must be taken to feed them into the conduit without tangling, and without rubbing against the side of the draw-in box. A small amount of slack cable should be left at each draw-in box and terminating box.

4.4 Aluminium Conduit

Aluminium conduit can be used in much the same way as steel conduit although more care must be taken in bending, owing to the tendency for the conduit to flatten. When threading, mineral oil or paraffin should be used as a lubricant. The fittings used are generally of cast or pressed aluminium alloy, galvanized iron or steel fittings can, however be used if desired. The aluminium must not be allowed to come into contact with copper or brass work as this results in corrosion; and in damp situations a coat of bituminous paint should be applied to reduce corrosion troubles. The main advantages of aluminium conduit are:

(i) It is light and easy to handle.
(ii) It is less liable to corrosion than steel.
(iii) It gives good earth continuity.
(iv) It is non-magnetic.

Its main disadvantage is that it is not so strong as steel, so extra protection may be needed where mechanical damage is likely.

4.5 Copper Conduit

This is a comparatively expensive system, which is sometimes used where prevention of corrosion is of first importance. The fittings used should be of bronze and may be either screwed or soldered to the conduit.

4.6 Flexible Conduits

Flexible conduit provides a useful means of protecting the final connections to motors, etc., where vibration or the possible need subsequently to adjust the position of the motor makes a rigid conduit connection unsatisfactory. In some situations the use of flexible conduit may avoid the need for complicated bends and sets when installing a machine. Flexible conduit should be used only for short runs where mechanical damage is unlikely to occur. Flexible conduit usually consists of a spirally wound, partially interlocked, light gauge galvanized steel strip, and may be either watertight or non-watertight, it can be obtained with a P.V.C. sheath if desired. A separate earth continuity conductor is necessary as the flexible conduit itself does not provide efficient earth continuity. Special brass adaptors are used to join the flexible conduit to ordinary conduit, one form being illustrated in Fig. 4.9.

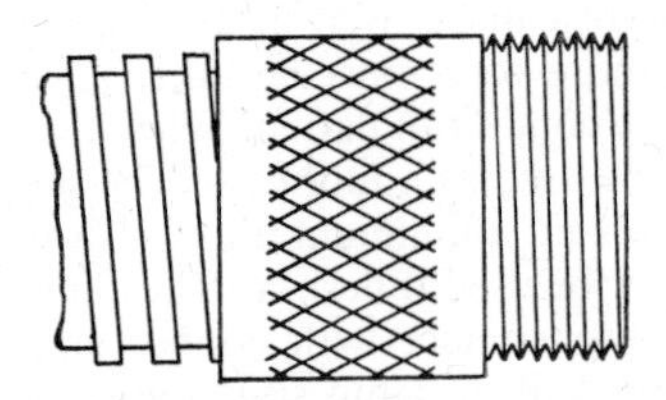

FIG. 4.9 Brass Adaptor for Flexible Conduit

4.7 Non-metallic Conduit

Both flexible and rigid grades of non-metallic conduit are available in the same diameters as steel conduits. The flexible grade is supplied in 50-ft. coils. Thus while the rigid type requires less fixings, the flexible type has the advantage that for most runs it can be installed without intermediate joints, and can be more easily set round obstacles. The flexible type can be easily bent without tools, it is advisable to warm the conduit slightly before bending especially in cold weather. The rigid type can be bent using the same techniques as for steel conduit, once again it is advisable to warm the conduit using a gentle blowlamp flame. Lengths of non-metallic conduit can be joined using a sleeve of the next larger size conduit. The conduit can be threaded for connection to fittings using ordinary electricians stocks and dies although most types will cut their own thread, Bostick adhesive may be used for sealing where necessary.

4.8 Cable Trunking

Where a large number of cables have to be installed or where the cable sizes themselves are large it is often preferable to use cable trunking rather than conduit. Trunking is available in sizes ranging from $1\frac{1}{2}$ in. by $1\frac{1}{2}$ in. to 12 in. by 6 in. in cross-section and in lengths of 6 or 8 ft. As this is a very rigid system (the trunking cannot be bent or set around obstacles) care must be taken in the planning stage to ensure that runs are as straight and direct as possible. The manufacturers provide a wide range of fittings to enable bends to be negotiated, some typical examples being shown in Fig. 4.10 (a–c).

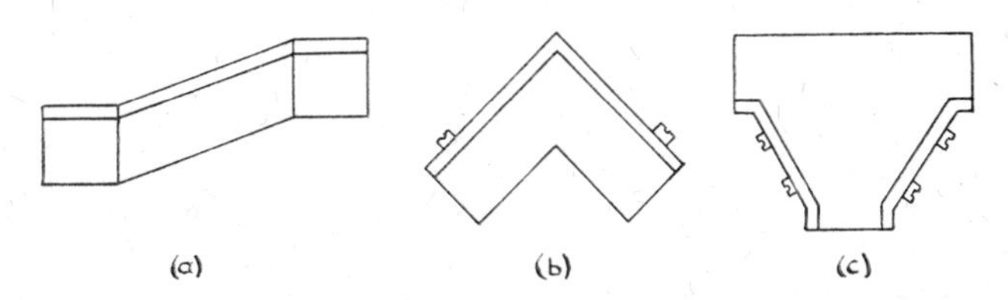

FIG. 4.10 (a–c) Trunking Fittings
(a) Off set; (b) 90° Angle bend; (c) Tee branch.

4.9 Bus-bar Trunking Systems

(*a*) Wiring systems which consist of copper bus-bars supported on insulators and enclosed in steel trunking are very convenient for use where large currents have to be handled. Two typical uses of this type of system are:

(i) Vertical rising mains.
(ii) Overhead bus-bar system.

(*b*) A vertical rising main is often employed in a multi-storeyed building to carry the supply to each floor. The incoming supply is connected to the bottom of the rising main, and a suitable tap off point is provided at each floor level. (Fig. 4.11.) A common type of trunking used for this purpose consists of 300A copper bus-bars enclosed in sheet steel trunking this is often supplied in sections of 10 to 12 ft. Since in the event of a fire hot gases and flames would tend to travel up the trunking due to convection it is vitally important that fire proof barriers are fitted at each floor level. A typical fire barrier consists of a 2 to 3 in. thick layer of glass wool supported by a hard asbestos board sheet and completely fills the trunking. Where the bus-bars pass through this barrier they must be fitted with insulating sleeves.

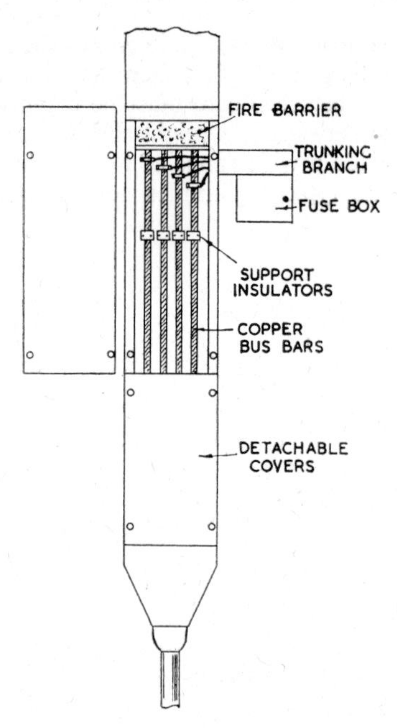

FIG. 4.11 Vertical Rising Main Bus-bar Trunking

(*c*) An overhead bus-bar system provides a convenient method of distribution in a factory or large workshop. A typical system used to provide a three-phase supply, has 'plug-in' tap off points at 3-ft. intervals into which a fused 'tap-off' box can be fitted. Thus it becomes easy to provide a supply position wherever it is required, and the position of the supply points can be readily changed if this should become necessary. Among the many types available are systems in which lighting fittings can be attached at any required point, and systems in which 13A socket outlets can be quickly fitted wherever they are required. Although the initial cost of installing an overhead bus-bar system is quite high, the fact that the system can be installed if necessary before the final positions of machines have been decided, and that the position of any machine may be readily altered at a later date without any expensive alterations to the wiring system, makes this system extremely convenient for industrial use. If long runs of bus-bar trunking are installed, it is necessary to provide 'expansion joints' at approximately 100-ft. intervals, to allow for expansion and contraction due to changes in temperature. Fire barriers must be installed at suitable intervals, particularly where the trunking passes through walls or partitions, to prevent the possible spread of a fire along the trunking. It must be noted that no other conductors must be installed in a trunking containing bare bus-bars, and that all lids and covers must be kept in position at all times, both to guard against accidental contact and to provide protection against vermin.

4.10 Under-floor Ducts

(*a*) In large buildings it is often an advantage to install a network of ducts in the solid concrete floors during the erection of the building, the ducts being subsequently used to accommodate the wiring of the electrical circuits. The main advantages of this type of system are:

(i) The major part of the accommodation for the wiring can be conveniently installed at an early stage during the construction of the building, even though the exact positions of the outlets may not be known at that time.
(ii) A concealed wiring system is provided, without the need for excessively long 'chases' in the concrete floors.
(iii) It is a relatively easy matter to alter the positions of outlets should this become necessary, e.g. owing to a change of tenancy.

(*b*) The duct system often takes the form of an intersecting grid so that every part of the floor area is within reasonable distance of a duct. Junction boxes are provided at intersections and at right-angle bends, the ducts themselves running in straight lines between the junction boxes. Several ducts may be laid side by side to accommodate the various services such as lighting, heating and telephones, etc., and in this case the junction boxes must be provided with a suitable means of effectively segregating the various services.

EXERCISES

The exercises which conclude this chapter are mainly intended to give practice in the use of heavy gauge steel conduit.

EXERCISE No. 4.1 **Running coupler.**

Materials

2 Short lengths $\frac{3}{4}$-in. heavy gauge steel conduit.
1 $\frac{3}{4}$-in. Socket.
1 $\frac{3}{4}$-in. Locknut.

Diagrams

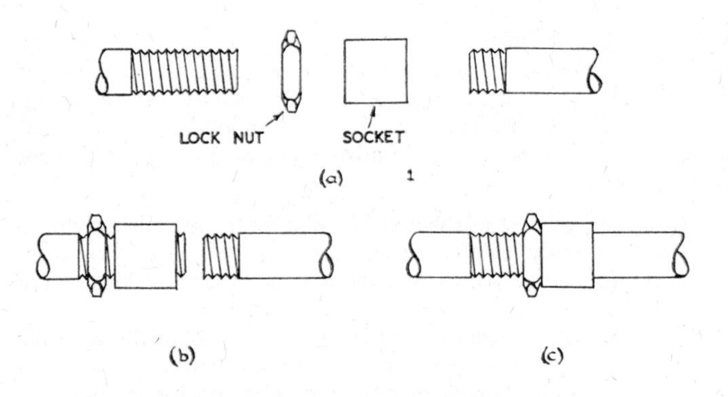

FIG. 4.12 (a–c) Exercise 4.1
(a) Step 1; (b) Step 2; (c) Step 3.

Procedure

1. Thread one length of conduit for one half the length of the socket and the other sufficiently to accommodate the locknut and the full length of the socket. Remove all burrs. (Fig. 4.12 (a–c).)
2. Screw first the locknut, then the socket, fully on to the length of conduit with the longer thread.
3. Butt the ends of the conduits together, and unscrew the socket from the first length of conduit so that it screws fully on to the second length of conduit. Tighten the locknut up to the socket.

Answer the following questions

1. Where is this type of coupling used?
2. Why is it important to ensure that the socket is tightened firmly on to the second length of conduit, and that the locknut is tightened firmly up to the socket?

EXERCISE No. 4.2 **To terminate conduit at a clearance entry.**

Materials

$\frac{3}{4}$-in. Heavy gauge steel conduit.
$\frac{3}{4}$-in. Socket.
$\frac{3}{4}$-in. Locknut.
$\frac{3}{4}$-in. Brass male bush.
$\frac{3}{4}$-in. Brass female bush.
Adaptable box with $\frac{3}{4}$-in. clearance hole.

Diagrams

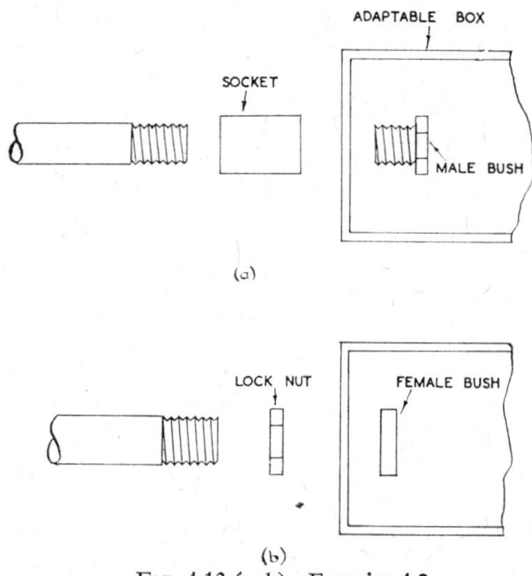

FIG. 4.13 (a–b) Exercise 4.2
(a) Method A; (b) Method B.

Procedure

METHOD A: Fig. 4.13 (a)

1. Thread the end of the conduit, sufficient to screw on the socket for half its length. Remove all burrs.
2. Screw on the socket.
3. Place the socket against the hole in the box and screw the male bush into the socket from inside the box.

METHOD B: Fig. 4.13 (b)

1. Thread the end of the conduit, sufficient to accommodate the locknut, the thickness of the adaptable box and the female bush, remove all burrs. Screw the locknut fully on to the conduit.
2. Pass the conduit through the hole in the box, and fit the female bush to the end of the conduit.
3. Tighten the locknut back against the side of the box.

Answer the following questions

What are the requirements of the I.E.E. Regulations with regard to:

(*a*) The enclosure of cables at conduit terminations?

(*b*) Ventilation of conduits?

EXERCISE No. 4.3 — One lamp with single-way switch control.

Materials

$\frac{3}{4}$-in. Heavy gauge steel conduit.
$\frac{3}{4}$-in. Tee box with lid.
$\frac{3}{4}$-in. Terminal box with ceiling rose.
$\frac{3}{4}$-in. Switch box with one single-way switch.
$\frac{3}{4}$-in. Female bush.
$\frac{3}{4}$-in. Spacer bar saddles.
3/·029 Red V.R.I.
3/·029 Black V.R.I.
B.C. lamp holder.
14/·0076 Flexible cord.

Layout

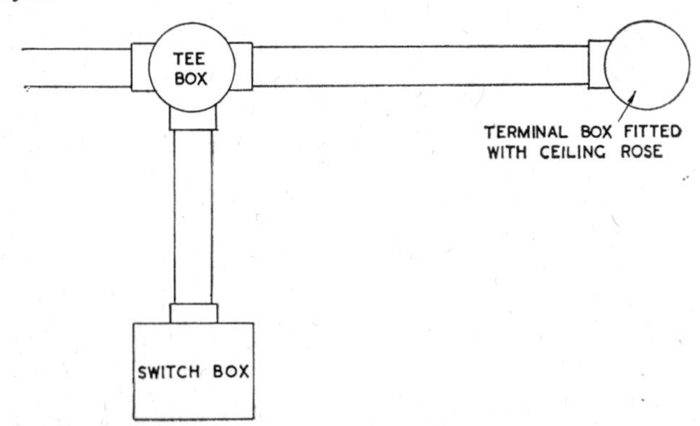

FIG. 4.14 Exercise 4.3

Procedure

1. Cut the conduit to the required lengths, thread the ends and remove all burrs.
2. Assemble the conduit and fittings as shown in Fig. 4.14.
3. Wire the completed conduit assembly to give one-way switch control of the lamp, leaving the supply ends of the wiring projecting a short distance from the supply position.

Answer the following questions

1. Why is it desirable to completely erect the conduit system before wiring?
2. Why is it good practice to use inspection tee boxes rather than inspection tees?

EXERCISE No. 4.4 — One lamp with single switch control and one lamp with two-way and intermediate switch control, using steel conduit.

Materials

$\frac{3}{4}$-in. Heavy gauge steel conduit.
1 $\frac{3}{4}$-in. Tee box with lid.
2 $\frac{3}{4}$-in. Tee box with ceiling rose.
1 $\frac{3}{4}$-in. Switch box with single-way switch.
2 $\frac{3}{4}$-in. Switch box with two-way switches.
1 $\frac{3}{4}$-in. Switch box with intermediate switch.
3 $\frac{3}{4}$-in. Spacer bar saddles.
1 15A S.P. & N. Switch fuse.
1 $\frac{3}{4}$-in. Socket.
1 $\frac{3}{4}$-in. Brass male bush.
3/·029 Red V.R.I.
3/·029 Black V.R.I.
14/·0076 Flexible cord.
2 B.C. Lamp holders.

Layout

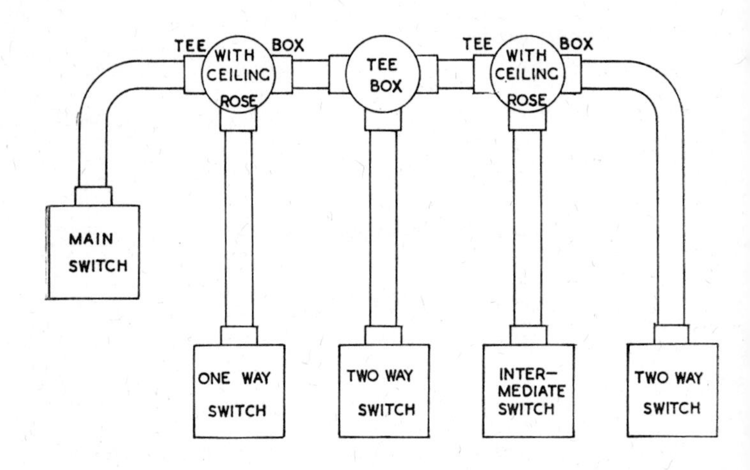

FIG. 4.15 Exercise 4.4

Procedure

1. Cut and bend the conduit to the required lengths, thread the ends and remove all burrs.
2. Assemble the conduit and fittings as shown in Fig. 4.15.
3. Wire the completed conduit assembly to give single-way control of one lamp and two-way and intermediate control of the other lamp.

Answer the following questions

1. Why do the I.E.E. Regulations require that the cables of a.c. systems installed in steel conduit shall always be so bunched that the cables of all phases and the neutral (if any) are drawn in to the same conduit?
2. What is the minimum permissible radius of bend for the conduit used in this exercise?

EXERCISE NO. 4.5

Master control circuit, using steel conduit.

Materials

¾-in. Heavy gauge steel conduit.
2 ¾-in. Through boxes.
2 ¾-in. Terminal boxes.
1 ¾-in. Angle box.
1 ¾-in. Inspection tee.
1 ¾-in. Brass female bush.
1 Ceiling rose.
1 B.C. lamp holder.
2 5A one-way lighting switches.
2 5A two-way lighting switches.
3/·029 Red V.R.I. cable.
3/·029 Black V.R.I. cable.
14/·0076 Flexible cord.

Layout

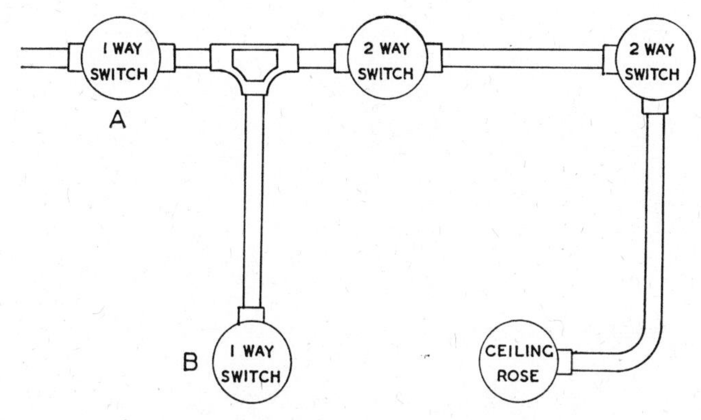

FIG. 4.16 Exercise 4.5

Procedure

1. Cut and bend the conduit to the required lengths, thread the ends and remove all burrs.
2. Assemble the conduit and fittings as shown in Fig. 4.16.
3. Wire the completed conduit assembly so that switches A and B provide 'master' control. Switch A must prevent the light from being switched on

by either of the two-way switches, and switch B must prevent the light from being switched off by either of the two-way switches.

Answer the following questions

What are the requirements of the I.E.E. Regulations with regard to:

(*a*) The situation of inspection boxes, draw-in boxes, etc.?
(*b*) Situations where conduits pass through floors, walls, ceilings, etc.?

EXERCISE No. 4.6 — To investigate a section of vertical rising main bus-bar trunking.

Apparatus

Section of vertical rising main bus bar trunking (fitted with **P.I.L.C.** end box, tap off unit, and fire barrier).

N.B. – It is desirable that this length of trunking is not connected to any supply.

Procedure

1. If the trunking to be investigated is connected to a supply make sure that it is switched off.
2. Make a detailed inspection of the trunking, paying particular attention to:
   (*a*) the method of connecting the incoming supply cable;
   (*b*) the method of connecting the tap off unit;
   (*c*) the method used to provide a fire barrier.
3. Make neat sketches (with leading dimensions shown) of:
   (*a*) the incoming cable arrangements;
   (*b*) the method of tapping off the supply from the vertical bus-bars;
   (*c*) the fire barrier.

Answer the following questions

1. For what purposes are vertical rising mains used?
2. Why is the fire barrier essential?

CHAPTER 5

Protection and Earthing

5.1 Protection

There are many ways of protecting both the installation and the user from the risk of electric shock or fire which may occur under fault conditions. In general a protective device is designed to disconnect the circuit whenever it detects a fault condition.

5.2 Shock Risk

A shock risk arises whenever accidental contact is made between the live conductor and exposed metal work. This risk can be guarded against either by efficient earthing or, where a good earth is not obtainable, by employing voltage operated, earth leakage, circuit breakers.

5.3 Fire Risk

Fire risks in electrical installations can arise in various ways which include:

(*a*) Sustained overloading of wiring or equipment.
(*b*) Faulty contacts or connections.
(*c*) Earth leakage currents.

(*a*) Every circuit must have some form of overcurrent protection, which often takes the form of a fuse, although miniature circuit-breakers are becoming increasingly popular. The overcurrent protection device must be capable of disconnecting the supply safely in the event of the most severe fault that can arise (i.e. a short-circuit across its own outgoing terminals). The current rating of a fuse or circuit breaker with inverse time characteristics, that is a circuit-breaker which operates in a time which is inversely proportional to the magnitude of the fault current, must not be more than the current rating of the cables protected by it. A circuit-breaker which operates instantaneously may have a rating up to twice that of the cables which it protects.

(*b*) Trouble due to faulty contacts and connections can only be avoided by the use of good quality accessories and fittings, and by using correct methods of installation. It should be realized that a loose or poor contact can attain a very high temperature even though only a comparatively small current is flowing, and the overcurrent device gives no protection under these circumstances.

(*c*) Earth leakage currents can give rise to fire risks, particularly if the earthing and bonding arrangements are not capable of carrying a sustained fault current without excessive heating. Once again danger often arises from inadequate connections, such as an earth clip which has worked loose, a poorly made conduit connection or corrosion of parts of the earthing system. It has been stated that the majority of fires attributed to electrical causes are associated with earth leakage conditions.

5.4 Fuses

A fuse provides the simplest form of overcurrent protection. The following terms are used in connection with fuses:

CURRENT RATING – This is the maximum current that a fuse will carry indefinitely without undue deterioration of the fuse element.

FUSING CURRENT – This is the minimum current that will 'blow' the fuse.

FUSING FACTOR – This is the ratio of the fusing current to the current rating,

$$\text{i.e. Fusing factor} = \frac{\text{Fusing current}}{\text{Current rating}}$$

(*a*) The rewirable fuse is a simple and cheap type of fuse which is still widely used despite its many disadvantages. It possesses only one real advantage, the cheapness of the fuse element. Its disadvantages are:

(i) It is too easy for an inexperienced person to replace the fuse element with wire of incorrect gauge or type.

(ii) Even when the correct wire is used for the fuse element the current rating and fusing factor may not be exactly as intended.

(iii) There is often undue deterioration of the fuse element due to oxidization.

(iv) A rewirable fuse is unsuitable for protecting circuits where very large fault currents may flow, as the fuse carrier is liable to disintegrate under these conditions.

(*b*) Cartridge fuses are an advance on rewirable fuses, as the rating of a replacement fuse element is determined by the manufacturer. Many fuse carriers are designed so that it is impossible to insert a fuse element of incorrect rating. Some advantages of cartridge fuses are:

(i) The rating is accurately known.
(ii) The fuse element is less prone to deteriorate.

Some disadvantages are:

(i) The fuse element is more expensive to replace than the rewirable type.
(ii) It is unsuitable for use where extremely high values of fault current may occur.

Both rewirable and cartridge fuses are widely used for protecting domestic installations and smaller industrial loads.

(*c*) The high rupturing capacity (H.R.C.) fuse has its characteristics carefully controlled by the manufacturer and, as its name suggests, can safely interrupt very large fault currents. These fuses are often used to protect large industrial loads, mains cables, and in other situations where very large fault currents can occur. Some advantages of H.R.C. fuses are:

(i) Its characteristics can be designed to suit the nature of the load.
(ii) The H.R.C. fuse is able to clear heavy fault currents safely.

The main disadvantage is that this type of fuse is more expensive than either the rewirable or cartridge type fuse.

5.5 Circuit-Breakers

Circuit-breakers may be used to disconnect automatically a faulty circuit. Miniature circuit-breakers (M.C.B.) have been developed in recent years, as an alternative to fuses, as a means of protection for domestic installations and other small loads. Some advantages are:

(*a*) The overload tripping characteristics are set by the manufacturer and cannot be altered.

(*b*) The characteristics are such that the circuit-breaker will trip for a small sustained overload but not on harmless transient overloads; the operation is instantaneous when a short-circuit occurs.

(*c*) Faulty circuits can be easily identified.

(*d*) Supply can be quickly and easily restored when the fault has been cleared.

5.6 Earthing

(*a*) The I.E.E. Regulations recommend that all circuits operating at a voltage exceeding extra low voltage shall be protected against dangerous earth leakage currents. This may be achieved by:

(i) Completely insulating all parts of the system,

or

(ii) Double-insulation of appliances,

or

(iii) Earthing of exposed metal apparatus, the earth terminals of socket outlets and all metal work associated with wiring systems, with certain exceptions such as cable clips, lamp caps, metal chains for suspending lighting fittings, etc.

(*b*) When a fault to earth occurs the fault current flows around the 'earth fault loop path' indicated in Fig. 5.1. The path taken by the current is from the live terminal, L, of the supply transformer along the live conductor to the fault, and then to the metal work affected. From the metal work the current flows via the earth continuity conductor (E.C.C.), earthing lead and consumer's earth electrode through the general mass of earth and so back to the neutral terminal, N, of the supply transformer. The earth fault loop path possesses impedance, i.e. it presents an opposition to the flow of alternating current which may be measured in ohms; the value of the fault current is limited by this impedance. If the impedance is low enough, sufficient current will flow to operate the overcurrent protective device. If it is not possible to obtain a low value of impedance then some form of earth leakage protection is required in addition to an overcurrent protective device.

5.7 Protective Multiple Earthing (P.M.E.)

(*a*) In some instances the method known as P.M.E. may be used to provide a low earth impedance. The principle of this system is that the earth continuity conductors are connected to the neutral service

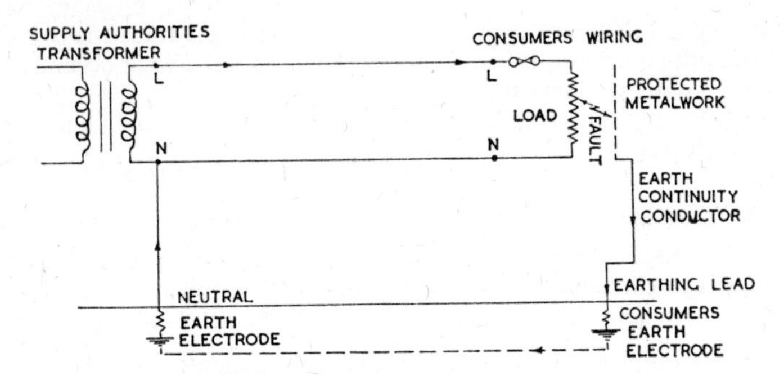

FIG. 5.1 Earth Fault Loop Path

conductor at the supply intake position, the neutral conductor being connected to earth at regular intervals as shown in Fig. 5.2.

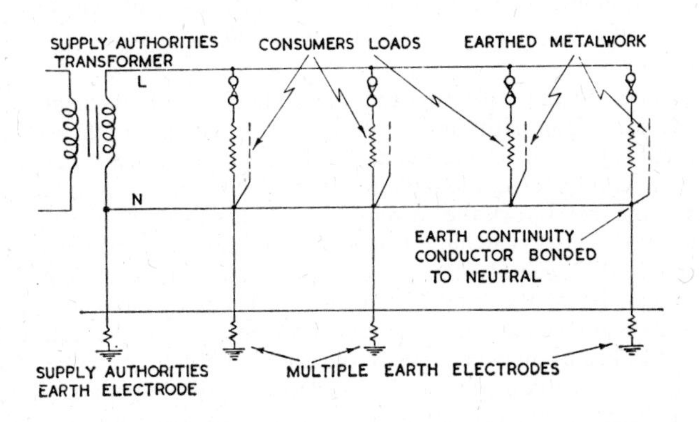

FIG. 5.2 P.M.E. System of Earthing

Some advantages of this system are:

(i) There is a low impedance path for fault current to flow via the neutral conductor.
(ii) The impedance to earth is low even though individual earth impedances may be high, since many earth connections are made in parallel.

Some disadvantages are:

(i) The shock risk that may arise if any earthed metal work associated with the system is not bonded to the neutral conductor or if the neutral conductor is broken.
(ii) Earth currents may circulate between the multiple earth electrodes causing interference with telephone systems.

(*b*) The following examples illustrate two dangerous conditions which could arise when P.M.E. is adopted.

(i) In the situation shown in Fig. 5.3(a) a fault has developed between the live conductor on a consumer's premises and some metal work connected to a good earth of resistance 1Ω, but which is not bonded to the neutral. The supply authority's earth resistance is 7Ω. The fault current will be $\frac{240}{7+1} = 30\text{A}$ which is not sufficient to blow the consumer's fuse, nevertheless the voltage drop across the supply authority's earth resistance is $7 \times 30 = 210\text{V}$. This means that the neutral conductor is raised to a potential of 210V above earth and so is all the 'earthed' metal work connected to it.
(ii) In the situation shown in Fig. 5.3(b) the neutral conductor has a break so the load current must return through the earth and, in so doing, causes voltage drops across both the consumer's earth impedance and the combined impedance of the earths connected to the remaining length of the neutral conductor. Thus for the situation shown, the load normally takes a current of $\frac{240}{10} = 24$ A. But, when the break in the neutral occurs, the load current is reduced to

$$\frac{240}{10 + 15 + 5} = 8 \text{ A}.$$

The voltage across the load will be reduced to $8 \times 10 = 80\text{V}$, the neutral conductor and all earthed metal work on the load side of the break is raised to a potential of $8 \times 15 = 120\text{V}$ above earth, while the neutral conductor and all 'earthed' metal work on the supply side of the break is raised to a potential of $8 \times 5 = 40\text{V}$ above earth.

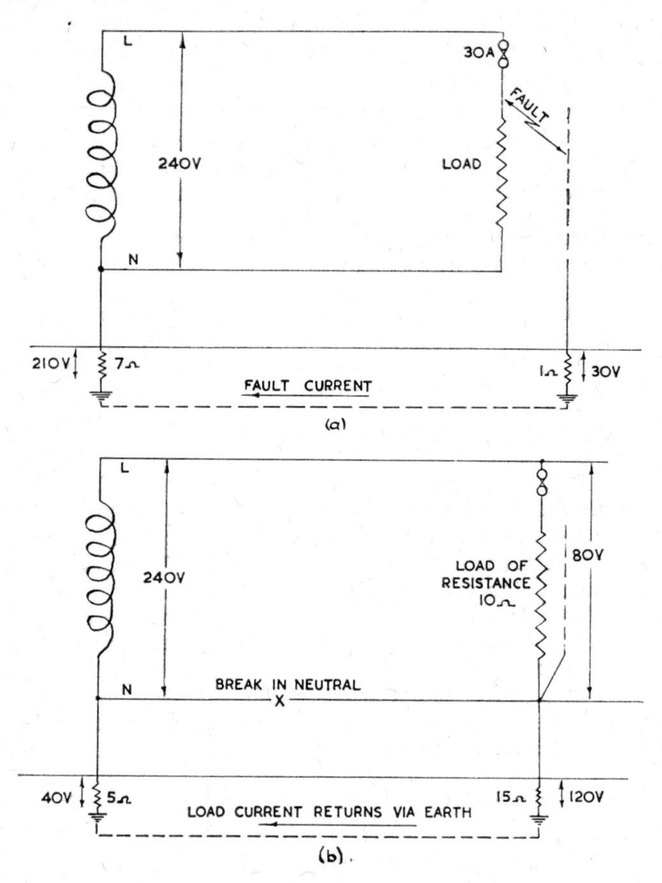

FIG. 5.3 (a–b) Dangerous Conditions when P.M.E. is in use
(a) Fault to earthed metal work not bonded to neutral; (b) Broken neutral conductor.

(*c*) Because of the inherent dangers which can arise with this system it is essential that the I.E.E. Regulations concerning it are strictly observed. The requirements are listed in Appendix 'F' of the I.E.E. Regulations, the more important points being:

(i) The use of this system must be authorized by the Minister of Power with the concurrence of the Post Master General (Regulation 4, Electricity Supply Regulations 1937).

(ii) All premises supplied must be connected so as to use the P.M.E. system of earthing.

(iii) The maximum resistance from neutral to earth must not exceed 10Ω.

(iv) No fusible cut out, automatic circuit-breaker, removable link or single pole switch shall be included in the neutral conductor.

(v) Earth continuity conductors must have a resistance of less than ½ ohm which must be tested by passing a current of not less than 10 amperes for a period of at least five minutes.

In general the above requirements limit the application of P.M.E. to new networks, since it is usually very expensive to adapt all the installations connected to an existing network to meet all the necessary requirements.

5.8 Earth Leakage Circuit-Breakers (E.L.C.B.)

In situations where a direct earth connection of low enough impedance cannot be obtained some form of earth leakage circuit-breaker should be installed. Earth leakage circuit-breakers fall into two classes, voltage and current operated types.

(*a*) Voltage-operated types (Fig. 5.4), are actuated by a rise in potential of the protected metal work caused by earth leakage currents. When an earth leakage current flows the potential of the protected metal work will rise because of the impedance of the earth fault loop path. This potential causes a current to flow through the trip coil of the E.L.C.B. and so to earth via the reference earth electrode. The maximum potential required to actuate the E.L.C.B. must not exceed 40V r.m.s. so that the danger of shock is reduced to a minimum.

(*b*) Current-operated types (Fig. 5.5), are actuated by excessive earth leakage current (15% of the rated current for the circuit or 5A whichever is greater). As is shown in Fig. 5.5, there is no direct connection between the earthing lead and the E.L.C.B., a 'core balance' transformer being used to detect the presence of earth leakage currents. Under normal conditions the current flowing in the live conductor equals the current returning in the neutral conductor. The two primary coils of the transformer are wound with opposite

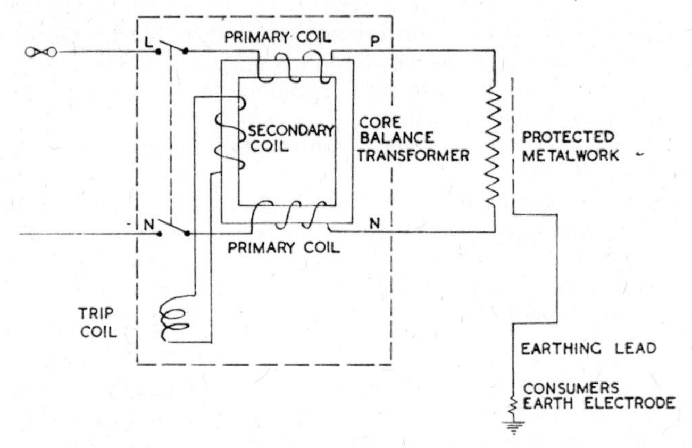

FIG. 5.4 Current Operated E.L.C.B.

polarity so, under normal conditions, their magnetic effects balance each other and no voltage is induced in the secondary coil. When an earth fault occurs the fault current returns via the earth loop path instead of the neutral conductor; thus the current taken from the

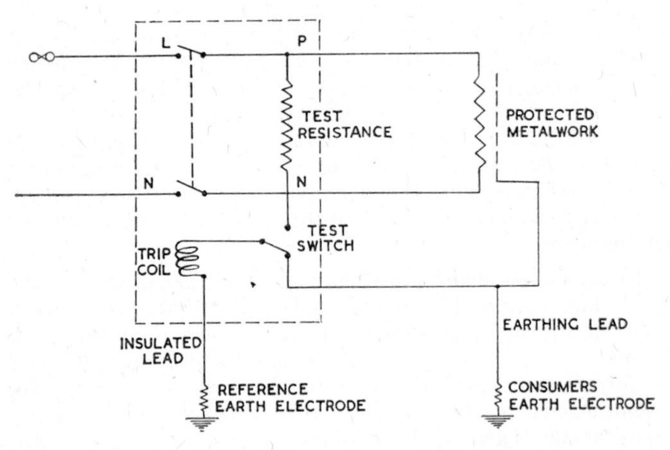

FIG. 5.5 Voltage Operated E.L.C.B.

live conductor exceeds the current returned to the neutral conductor. This disturbs the balance between the magnetic effects of the two primary coils and a magnetic flux is set up in the core thus inducing a voltage in the secondary coil which, in turn, operates the tripping relay.

(*c*) Voltage-operated E.L.C.B.s are used in domestic and smaller industrial installations to give protection against shock risk where a good earth cannot be obtained. Current-operated E.L.C.B.s are useful in larger installations where very heavy fault currents would flow before the overcurrent device operated, thereby creating a serious fire risk.

EXERCISES

The exercises which conclude this chapter are designed to illustrate various methods of circuit protection.

EXERCISE No. 5.1 — To compare the operation of a rewirable fuse with a miniature circuit-breaker.

Apparatus

Demonstration board No. 2.
0–20A a.c. ammeter.
Loading panel.

Diagram

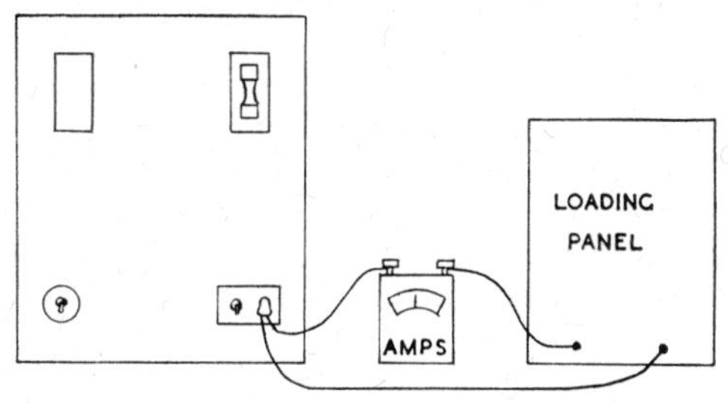

FIG. 5.6 Exercise 5.1

Procedure

1. Wire the fuse carrier with 5A rated wire.
2. Connect demonstration board, loading panel and ammeter, as shown in Fig. 5.6, and plug in to the mains supply.
3. Set changeover switch to 'M.C.B.' position, switch on 20A of load current and note that M.C.B. trips after a short time.
4. Set changeover switch to 'fuse' position, and note that the fuse blows immediately.
5. Disconnect demonstration board from the mains supply and replace the fuse wire.
6. Switch off load and connect demonstration board to the supply.
7. Set changeover switch to 'M.C.B.' and switch on 8·5A of load current and note that the M.C.B. takes a long time to trip.
8. Set changeover switch to 'fuse' and note that the fuse blows after a few seconds.
9. Disconnect demonstration board from the supply, and replace fuse wire.
10. Switch off load and connect demonstration board to the supply.
11. Set changeover switch to 'M.C.B.' and switch on 5A of load current and note that the M.C.B. does not trip even when the load is sustained for a long time.

12. Set changeover switch to 'fuse' and note that the fuse does not blow.

Answer the following questions

1. What are the advantages and disadvantages possessed by a rewirable type fuse?
2. Explain the terms current rating, fusing factor and fusing current as applied to fuses.
3. Discuss the advantages and disadvantages of automatic circuit breakers as compared with fuses.

EXERCISE No. 5.2 **Investigation of an earth fault.**

Apparatus

Demonstration board No. 3.
Ammeter 0–1A a.c.
Voltmeter 0–3V a.c.

Diagram

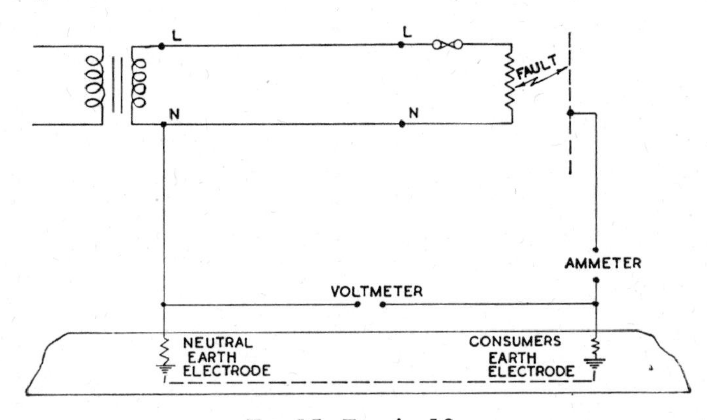

FIG. 5.7 Exercise 5.2

Procedure

1. Connect ammeter and voltmeter.
2. Ensure that the rheostat is at its lowest setting and plug in to the mains supply.
3. Adjust the rheostat until a fault current of 1A is indicated on the ammeter.
4. Note the potential difference between neutral and earth which is indicated by the voltmeter.

Answer the following questions

1. What value of impedance is encountered by the fault current.
2. What would be the value of fault current on this installation if a short-circuit occurred between the live conductor and earth, and the supply transformer provided a voltage of 240V?

3. Draw a neat circuit diagram showing the path taken by the fault current.
4. What value of test current is required by the I.E.E. Regulations for measuring the Earth Fault Loop Impedance?
5. What are the basic requirements of earthing as laid down by the I.E.E. Regulations?

EXERCISE No. 5.3 — Protective multiple earthing (comparison with direct earthing).

Apparatus

Demonstration board No. 4.
Voltmeter.
Ammeter.

Diagram

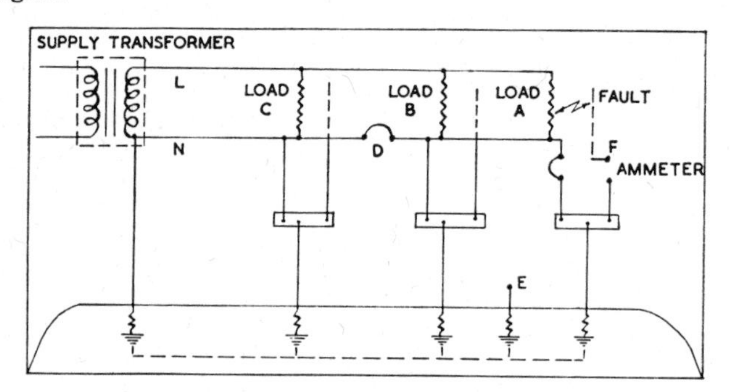

FIG. 5.8 Exercise 5.3

Procedure

1. Adjust the rheostat to its lowest setting, and make sure that the neutral link at D is in place.
2. Connect the ammeter to register the fault current at load A. Connect the voltmeter to register the potential between the general mass of earth (terminal E), and the earth bar at load A.
3. Disconnect the earth bar from the neutral conductor at load A thus giving a simple direct earthing arrangement.
4. Connect the demonstration board to the supply, and adjust the rheostat to give a suitable fault current.
5. Note the ammeter and voltmeter readings and calculate the impedance offered to the flow of fault current:

$$\text{Impedance} = \frac{\text{Voltmeter reading}}{\text{Ammeter reading}}$$

6. Disconnect the demonstration board from the supply and reconnect the earth bar to the neutral conductor at load A thus producing a protective multiple earthing arrangement.

7. Reconnect the demonstration board to the supply and adjust the rheostat to give a suitable fault current.
8. Note the ammeter and voltmeter readings and again calculate the impedance offered to the flow of fault current.

Answer the following questions

1. What is meant by Protective Multiple Earthing?
2. What is the principal advantage of protective multiple earthing?

EXERCISE No. 5.4

Protective multiple earthing (fault to earthed metal which is not bonded to neutral).

Apparatus

Demonstration board No. 4.
3 Voltmeters.
Ammeter.

Diagram

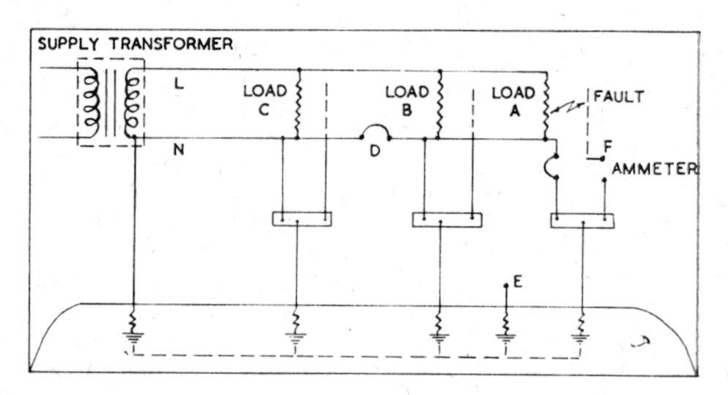

FIG. 5.9 Exercise 5.4

Procedure

1. Adjust the rheostat to its lowest setting, making sure that the neutral link at D is in place and that the earth bar at load A is connected to the neutral.
2. Connect the ammeter between terminals F and E so as to register the fault current flowing directly to earth. Connect the voltmeters to register the potentials between the general mass of earth (terminal E) and the earth bars at loads A, B and C.
3. Connect the demonstration board to the supply and adjust the rheostat to give a suitable fault current.
4. Note all voltmeter readings.

Answer the following questions

1. Explain why a fault current flowing directly to earth causes a rise in potential of all the earth bars.
2. Why is the use of protective multiple earthing usually restricted to installations served by new mains networks?

EXERCISE No. 5.5 **Protective multiple earthing (broken neutral).**

Apparatus

Demonstration board No. 4.
3 Voltmeters.

Diagram

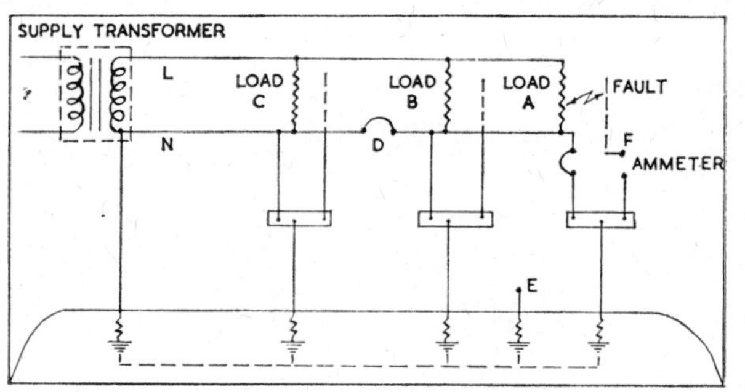

FIG. 5.10 Exercise 5.5

Procedure

1. Make sure that the neutral link at D is in place and that the earth bar at load A is connected to the neutral.
2. Connect the voltmeters to register the potentials between the general mass of earth (terminal E) and the earth bars at loads A, B and C.
3. Connect the demonstration board to the supply.
4. Note all the voltmeter readings.
5. Disconnect the demonstration board from the supply, and disconnect the neutral link at D.
6. Reconnect the demonstration board to the supply and again note the voltmeter readings.

Answer the following questions

1. Explain why a break in the neutral conductor causes a rise in potential at all the earth bars.
2. Discuss in your own words the advantages and disadvantages of protective multiple earthing.

EXERCISE No. 5.6

To connect a voltage-operated earth leakage circuit-breaker.

Apparatus

Voltage-operated earth leakage circuit-breaker.
Main switch.
13A Socket outlet.
Earth bar.
Reference earth electrode (see note).

Diagram

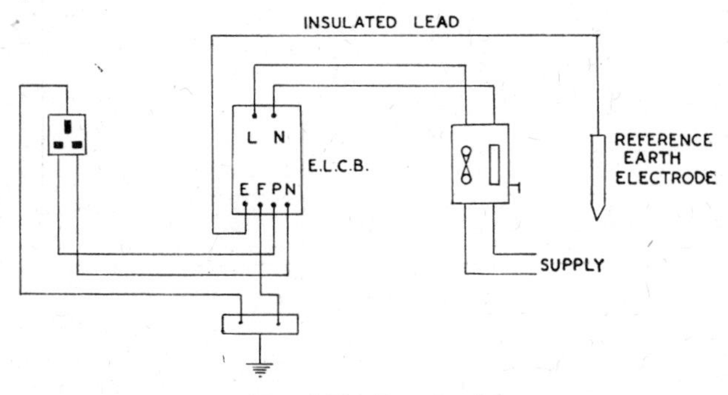

FIG. 5.11 Exercise 5.6

Note. – In order to simulate the conditions in which an earth leakage circuit-breaker might well be used the consumer's earth bar is permanently connected to a good earth via a 5-ohm resistor. The reference earth electrode is represented by a short length of copper earth rod which is directly connected to a good earth. These connections are not shown in Fig. 5.11.

Method

1. Connect apparatus in accordance with Fig. 5.11.
2. Switch on and press the test button to check that the earth leakage circuit-breaker is operating correctly.

Answer the following questions

1. Why is the connecting lead from the earth leakage circuit-breaker to the reference earth electrode insulated?

2. Why should the reference earth electrode be outside the effective resistance area of the main installation earth electrode?
3. Is this type of earth leakage circuit-breaker concerned with the potential of the metal work it is protecting, or with the supply voltage?
4. With the aid of a carefully drawn circuit diagram explain the operation of a voltage-operated earth leakage circuit-breaker.

EXERCISE No. 5.7 **To investigate the operation of a voltage-operated earth leakage circuit-breaker.**

Apparatus

Demonstration board No. 5 (artificial load with earth fault).
Voltage-operated earth leakage circuit-breaker.
Voltmeter 0–20V a.c.
Main switch.
Earth bar.

The main switch and earth bar should be mounted in a convenient position in the workshop and permanently connected to the necessary supplies.

Diagram

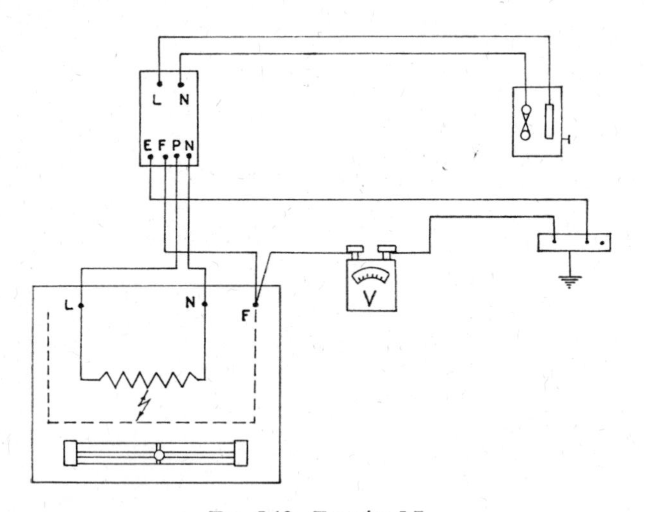

FIG. 5.12 Exercise 5.7

Procedure

1. Fix the earth leakage circuit-breaker in a suitable position near the main switch and earth bar.
2. Ensure the main switch is off.
3. Connect the circuit as shown in Fig. 5.12.
4. Set the slider on the rheostat to the neutral end.

5. Close the circuit-breaker and then switch on the supply.
6. Slowly adjust the slider on the rheostat until the earth leakage circuit-breaker trips, noting the voltmeter reading when this occurs.

Answer the following questions

1. What is tripping voltage for the earth leakage circuit-breaker tested?
2. What is the maximum value of tripping voltage permitted by the I.E.E. Regulations for this type of earth leakage circuit-breaker?
3. Does the protection afforded by this type of earth leakage circuit-breaker guard against shock risk or fire risk?
4. Why is it necessary to include fuses in the circuit even though an earth leakage circuit-breaker has been fitted?

EXERCISE No. 5.8

To investigate the operation of a current-operated earth leakage circuit-breaker.

Apparatus

Demonstration board No. 4 (artificial load with earth fault).
Current-operated earth leakage circuit-breaker (incorporating a core balance transformer).
Ammeter 0–3A a.c.
Main switch and earth as used for Exercise No. 5.7.

Diagram

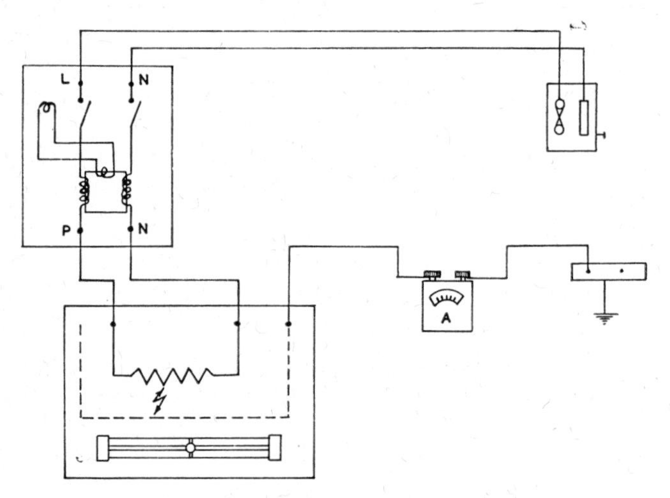

FIG. 5.13 Exercise 5.8

Procedure

1. Ensure the main switch is off.
2. Fix the earth leakage circuit-breaker in a suitable position near the main switch and earth bar.
3. Connect the circuit as shown in Fig. 5.13.
4. Set the slider on the rheostat to the neutral end.
5. Close the circuit-breaker and then switch on the main supply.
6. Slowly adjust the slider on the rheostat until the earth leakage circuit-breaker trips, noting the ammeter reading when this occurs.

Answer the following questions

1. Explain how an earth fault produces the out of balance condition needed to operate this type of earth leakage circuit-breaker.
2. Where is this type of earth leakage circuit-breaker mainly used?
3. With the aid of a carefully drawn circuit diagram explain the action of a current-operated earth leakage circuit-breaker paying special attention to the part played by the core balance transformer.

CHAPTER 6

Tests on Completed Installations

6.1

Every new installation, or extension to an existing installation, should be thoroughly tested before it is permanently connected to the supply, in order to be sure that it will function correctly and safely. It is also desirable to inspect and test the installation at regular intervals so that any serious deterioration of cables, fittings or earthing arrangements can be detected before serious danger arises. The I.E.E. Regulations recommend that tests shall be carried out in the sequence:

(*a*) verification of polarity,
(*b*) insulation tests,
(*c*) earthing tests.

6.2 Polarity Tests

(*a*) Tests should be made to ensure that the following requirements are satisfied.

(i) All single-pole switches should be wired in the non-earthed ('live') conductor.
(ii) The outer, or screwed, contacts of Edison screw-type lamp holders should be connected to the neutral conductor.
(iii) Socket outlets should be tested to ensure that the phase conductor is connected to the terminal marked 'L', the neutral conductor is connected to the terminal marked 'N', and the earth continuity conductor is connected to the terminal marked 'E'.

(*b*) Polarity tests can be made before the circuit is connected to the supply using either a low reading ohmmeter or a bell and battery. The method employed is to check the continuity of each conductor from the main switch, or distribution fuse board, to the switch or socket outlet concerned.

(*c*) Polarity tests can be made after the circuit is connected to the supply using either a test lamp, a neon voltage indicator or a voltmeter. These methods are limited in that they will only indicate which conductor is 'live', as no distinction can be made between the earth and neutral conductors. When carrying out tests on a live circuit great care should be exercised because of the risk of shock. Only properly constructed test lamps should be used, with the leads fitted with insulated test prods.

6.3 Insulation Tests

(*a*) The I.E.E. Regulations recommend that two insulation resistance tests should be made.

(i) An insulation resistance test must be made between earth and all non-earthed conductors connected together. This ensures that there are no faults to earth from either the 'live' or neutral conductors.

(ii) A further insulation test should be made between the 'live' and neutral conductors to ensure that there are no short-circuits in the wiring.

(*b*) Since every outlet (switch, lighting fitting, socket outlet, etc.) is a potential source of leakage the insulation resistance value to be expected is inversely proportional to the number of outlets. The I.E.E. Regulations state that the minimum acceptable value is

$$\frac{50}{\text{No. of outlets}}\ \text{megohms, or}\ \frac{12{\cdot}5}{\text{No. of outlets}}\ \text{megohms}$$

where P.V.C. insulated cables are used.

(*c*) The testing voltage used should be a d.c. voltage of at least twice the supply value except that it need not exceed 500V. If the value obtained is less than 0·5 megohms most testing instruments will be supplying a greater leakage current than that for which they are designed and this will cause the testing voltage to be less than the rated value. Therefore, in these circumstances, the wiring under test should be subdivided and each section tested separately.

6.4 Continuity of Ring Circuit

The continuity of each conductor of a ring circuit should be tested to ensure that each conductor is in fact continuous throughout the ring.

6.5 Earthing Tests

Tests should be made to ensure that the earthing arrangements are satisfactory.

(*a*) The effectiveness of earthing of an a.c. installation can be checked using either:

(i) Line-earth loop tester, or
(ii) Neutral-earth loop tester (except where Protective Multiple Earthing is in use).

The current used for these tests should be 50 c/s a.c. and of the order of 1·5 times the rated current of the circuit but need not exceed 25A. The value of the loop impedance obtained from the test should be low enough to permit a current of 3 times the rating of the fuse or 1·5 times the setting of the overload circuit-breaker to flow in the event of an earth fault.

(*b*) Should it become necessary to make a separate test of the earth continuity conductor of an a.c. installation, it is preferable to use alternating current of 1·5 times the rating of the circuit subject to a maximum of 25A, and the measured impedance should not exceed 1 ohm. A hand-tester giving a small current may be used, in which case the measured resistance should not exceed $\frac{1}{2}$ ohm, if steel conduit forms any part of the earth path.

(*c*) Direct-current installations should be tested using a d.c. current which may be obtained from a secondary battery and rheostat, the value of current used being of the order of 1·5 times the rating of the circuit under test but not exceeding 25A; alternatively, a hand-tester may be used. The resistance value obtained from the test should not exceed 1 ohm.

6.6 Earth Electrode Resistance

An earth electrode can consist of a metal rod or rods, a system of underground metal water pipes, or some other conducting object, providing an effectual connection with the general mass of earth. A current flowing to earth must overcome not only the resistance of the conductors forming the earth electrode but also the resistance of the surrounding soil. As the current leaves the electrode it fans out; thus, as the current travels farther from the electrode, the effective cross-sectional area of the soil through which it flows is increased. The overall effect is to produce a graded resistance which is concentrated mainly in the soil immediately surrounding the electrode, the area containing virtually all the resistance being called the 'earth electrode resistance area'.

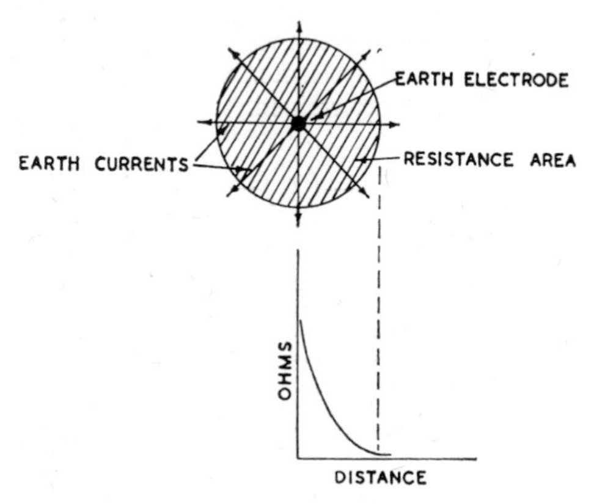

FIG. 6.1 Earth Electrode Resistance Area

6.7 Measurement of Earth Electrode Resistance

The basic method of measuring earth electrode resistance is to pass a current, I, from the earth electrode, E, to a current electrode, C, which is usually in the form of a copper spike. The resulting p.d. between the earth electrode and the general mass of earth is measured with the aid of a potential electrode, P. This arrangement is illustrated in Fig. 6.2. If P is close to E the voltage recorded will be too small, since P will be within the resistance area of E and not all of the electrode's earth resistance is included in the test. When P is placed outside the resistance area of E it is found that the same reading, V, is obtained no matter how far P is from E, until P is placed inside the resistance area of C, whereupon the voltage rises again due to the increased resistance. The earth electrode resistance is then given by Ohm's Law $\left(\text{i.e. } R = \frac{V}{I}\right)$.

If a constant value for V cannot be obtained for at least three positions of P between E and C the current electrode, C, is too close to the earth electrode, E, so that the resistance areas overlap. Electrode, C, should then be moved farther away until at least three constant readings can be obtained.

The simple method of earth electrode measurement described above is seldom used in practice, owing to the difficulties caused by electrolytic effects and by stray earth currents. Special earth testing instruments, which pass an alternating current between the electrodes are available; the design of these instruments compensates for the adverse effects of electrolysis, etc., and so enables the true earth electrode resistance to be determined.

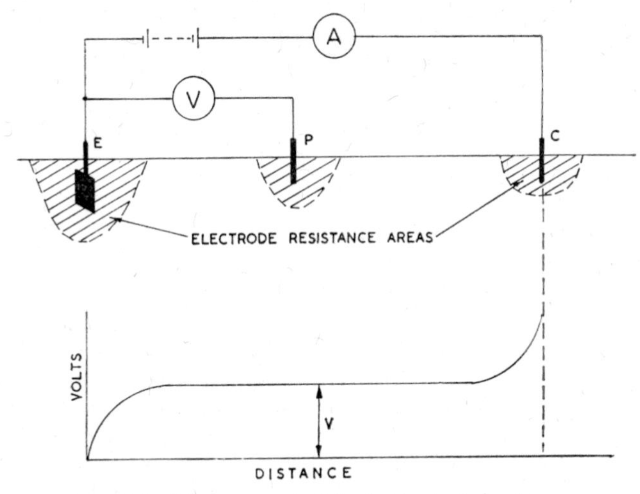

FIG. 6.2 Measurement of Earth Electrode Resistance

6.8 Fault Tracing

(*a*) The faults which may arise in an electrical installation can be grouped into three main classes:

(i) Open circuits.
(ii) Short-circuits between phase and neutral.
(iii) Earth faults.

When a fault is investigated a systematic method must be adopted, for a haphazard series of tests, carried out at random, is seldom likely to locate the trouble quickly. A preliminary investigation of the symptoms will sometimes make the cause of the fault obvious; for example, a report that sparks were seen coming from a lighting fitting after which the lights refused to function could very well lead to discovering quickly a faulty flexible cord. When the cause of the trouble is not obvious the first steps should always be aimed at narrowing the field to be investigated. A fault which affects only one sub-circuit almost certainly lies in that sub-circuit, while a fault affecting several sub-circuits may well be in the sub-main feeding the group of sub-circuits. When the faulty circuit is located, tests on individual conductors and fittings should quickly reveal the cause of the trouble.

(*b*) An open circuit can occur if a conductor is broken, or pulled out of a terminal, or if a switch or lamp holder, etc., fails to make efficient contact. The result of an open circuit in phase and neutral conductors is that the apparatus supplied by the circuit cannot operate. An open circuit in an earth continuity conductor gives rise to a shock risk if any earth leakage is also present. A continuity test, with the supply switched off, provides a good method of locating an open circuit.

(*c*) To test for a suspected open circuit in the lighting circuit wired with single core cables in conduit, as illustrated in Fig. 6.3, first switch off the supply and connect the switch feed to earth at its supply end. The lighting switch, S, should then be closed and the continuity tester connected between earth and terminals, A, B and C, in turn, using the conduit as a convenient earth return. A high reading indicates an open circuit conductor. For example, if a low reading is obtained at A but a high reading at C this indicates either a faulty switch or an open circuit in the switch wire. A further test between the switch terminals is sufficient to determine where the fault lies.

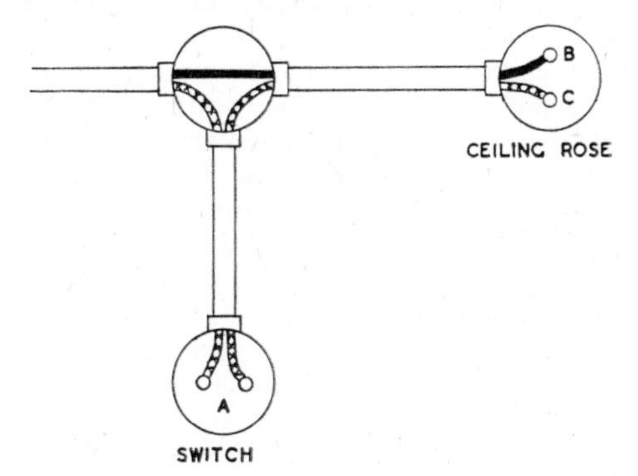

FIG. 6.3 Continuity Test (Screwed Conduit Installation)

(*d*) Fig. 6.4 shows a lighting circuit wired in all-insulated cables using joint boxes. As there is no convenient earth return, the following method may be adopted. Switch off the supply, close the lighting switch and make a good connection between the ceiling rose terminals. The continuity tester can then be connected between terminals D and E at the joint box to test the switch drop and switch, and between terminals E and F to test the cable feeding the lamp.

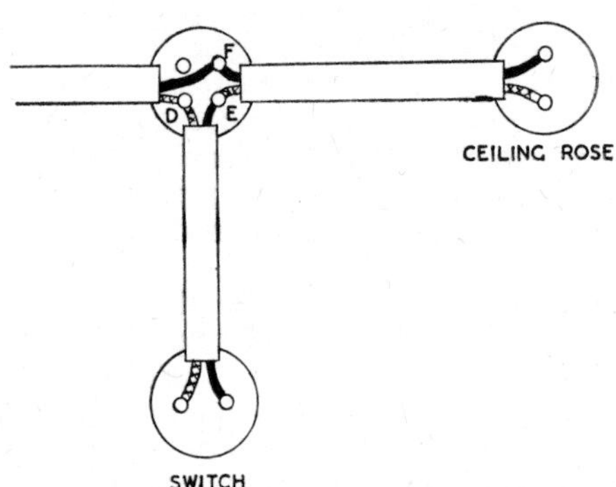

FIG. 6.4 Continuity Test (All-Insulated Wiring System)

Whenever tests are carried out on a circuit which involve a temporary alteration of the wiring, it is important to restore the circuit connections to their normal condition before attempting to restore the supply.

(*e*) A short-circuit between phase and neutral can occur as a result of damaged insulation, because of bad workmanship, leaving bare wires in joint boxes or fittings, or by conductors coming loose from terminals and subsequently moving so as to come in contact with the conductor of opposite polarity. The result is usually to blow the circuit fuses or trip the circuit-breaker; a short-circuit can result in the overheating of conductors and often causes sparking or arcing at the point where it occurs. To trace a short-circuit, first disconnect the phase and neutral conductors of the suspected circuit from the terminals of the fuse board or main switch. Switch off all switches in the circuit, and remove any lamps, etc. Test the insulation resistance between the phase and neutral conductors; if a high reading is obtained, close the switches one by one until the reading drops to zero, the fault is then located in the section of wiring controlled by the last switch to be closed. If a low, or zero, reading is obtained on the first test then the circuit will have to be subdivided by disconnecting at convenient points, until the faulty length of cable is isolated.

(*f*) An earth fault from the live conductor to a good earth will have similar effects to a short-circuit between phase and neutral conductors, resulting in fuses blowing, etc., and may be located in a similar manner to a short-circuit, the insulation tests being carried out between the phase conductor and earth. A leakage from the live conductor to metal work which is not soundly earthed may give rise to shocks and, in this event, both the insulation resistance and the earth continuity should be investigated.

(*g*) An earth fault on the neutral conductor is seldom discovered except as the result of routine insulation tests. Although this type of fault can persist in many cases without apparently affecting the operation of the circuit it is nevertheless important to rectify such faults when they are discovered. This is because, when such a fault occurs, currents may circulate in and around the earthing arrangements often giving rise to shock and fire risks.

EXERCISES

The exercises which conclude this chapter are designed to illustrate practical methods of performing the principal tests.

EXERCISE No. 6.1 **Polarity testing.**

Apparatus

Voltage indicator (with well shrouded test leads.
Low reading ohmmeter.
Demonstration board No. 6.

Diagrams

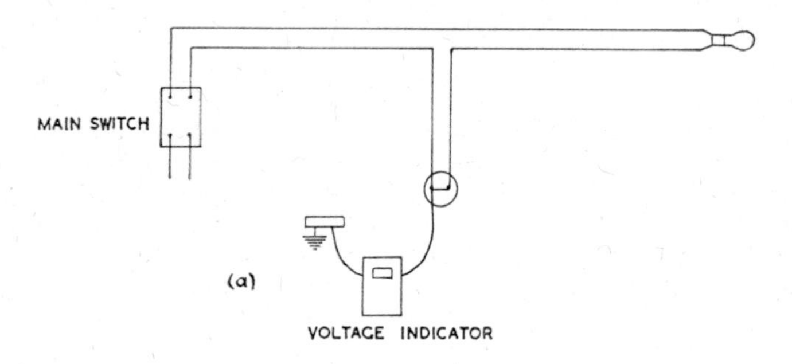

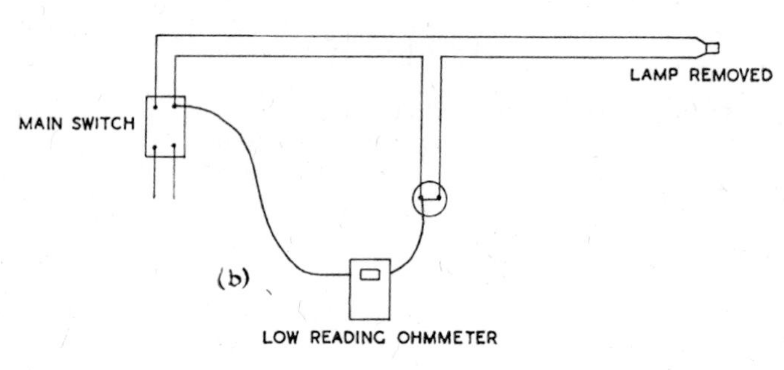

FIG. 6.5 (a–b) Exercise 6.1
(a) Method 1; (b) Method 2.

Procedure

METHOD 1

1. Remove cover from lighting switch, connect demonstration board to the supply.

2. Close main switch and lighting switch.
3. Connect the voltage indicator between earth and one pole of the lighting switch, taking great care not to make personal contact with the exposed live metal of the switch. Note that a voltage indication is obtained when the polarity is correct.
4. Operate the switch on the back of the demonstration board and repeat the test.

METHOD 2

1. Open main switch and remove fuses; close lighting switch.
2. Remove lamp and cover from lighting switch.
3. Connect one lead from the ohmmeter to the outgoing live terminal at the main switch and the other lead from the low-reading ohmmeter to one pole of the lighting switch. Note that a low reading indicates correct polarity.
4. Operate the switch on the back of the demonstration board and repeat the test.

Answer the following questions

1. Why are polarity tests important?
2. When using Method 1, why does a voltage indication show correct polarity?
3. When using Method 2, why does a low reading indicate correct polarity, and why is it essential to remove the lamp?

EXERCISE No. 6.2 **Circuit tracing.**

Apparatus

Low reading ohmmeter.
Test circuit.

Test Circuit

The test circuit is constructed using two adaptable conduit boxes linked by a suitable length of metal trunking. A five-way terminal block is fixed in each adaptable box and electrical connections made between the terminals at each end.

Diagram

FIG. 6.6 Exercise 6.2

Procedure

1. Connect terminal 1 to a good earth at box A.
2. Connect one lead from the low reading ohmmeter to a good earth at B. Connect the other ohmmeter lead in turn to each of the terminals, a, b, c, d, e, until a low reading is obtained, this then is the terminal, to which the wire from terminal 1 is connected.
3. Note the result of the test on a connection chart (that is if it is found that the wire from terminal 1 is connected to, say, terminal d then note it as 1——d).
4. Repeat the test for terminals 2 to 5 in order.

Connection Chart

Terminal Block A Connects to Terminal Block B

1
2
3
4
5

Answer the following questions

1. Explain, with the aid of a carefully drawn diagram, why a low reading is obtained only when the correct terminal is connected to the ohmmeter.
2. Give a practical case where it might become necessary to use this method of circuit tracing.

EXERCISE No. 6.3 **Insulation tests.**

Apparatus

Demonstration board No. 7.
Insulation resistance tester.

Diagram

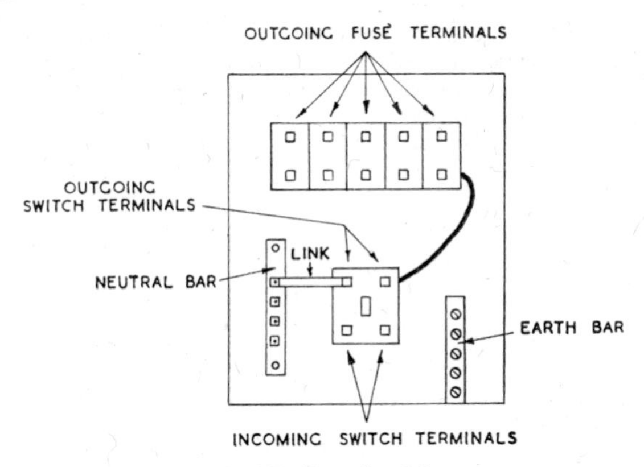

FIG. 6.7 Exercise 6.3

Procedure

N.B. - *In NO circumstances must any connections be made to the incoming terminals of the main switch.*

1. Switch off and remove cover from consumer's unit. Insert all fuses.
2. Link together the *outgoing* 'live' and 'neutral' terminals of the main switch (a short length of V.R.I. cable fitted with 'crocodile' clips provides a convenient means of doing this).
3. Connect the leads of the insulation tester, one to a good earth and the other to the 'live' outgoing terminal of the switch, operate the instrument and note the reading; this is the insulation resistance between conductors.
4. Remove the link between 'live' and 'neutral' terminals of the switch and connect the insulation tester leads one to the outgoing 'live' terminal and the other to the 'neutral' terminal of the main switch. Again operate the instrument and note the reading, which is the insulation resistance between conductors.

If either of the readings taken above is less than 0·5 megohm, then the circuits must be 'subdivided' for further tests. Proceed as follows:

5. Remove all fuses.
6. Join one lead of the insulation tester to earth and the other to the outgoing terminal of a fuse, also link the outgoing terminal under test to the neutral bar in the consumer's unit; operate the instrument and note the reading.
7. Test each circuit in turn as above, so that the insulation resistance to earth of each circuit is determined.
8. Join one lead from the test instrument to the neutral bar in the consumer's unit, and the other to an outgoing fuse terminal, operate the instrument, and note the reading.
9. Test each circuit in turn as above, so that the insulation resistance between conductors for each circuit is determined.

Results

Circuit	*Insulation Resistance to Earth*	*Insulation Resistance between Conductors*
Combined		
1		
2		
3		
4		
5		

Answer the following questions

1. Explain the difference between insulation resistance and conductor resistance.
2. When carrying out the test for insulation resistance between conductors, it is necessary to close all lighting switches and remove all lamps. Why is this?
3. What are the formulae for calculating the minimum acceptable insulation resistance quoted in the I.E.E. Regulations? Calculate the acceptable value for each circuit tested (using the data provided on the demonstration board). Do all the circuits tested comply with minimum standards?
4. What are the principal causes of low insulation resistance?

EXERCISE No. 6.4 Test of effectiveness of earthing.

Apparatus

Demonstration board No. 8.
Neutral earth loop tester.

Procedure

1. Examine the demonstration board and check that:
   (*a*) The artificial earth resistance on the demonstration board is set to zero.
   (*b*) The polarity of the socket is correct.
2. Connect the test instrument and note the reading.
3. Repeat the above test but with artificially increased resistance in the earth lead to represent different conditions.
4. From your results check the prospective earth fault current that could flow:

$$\text{i.e. Earth Fault Current} = \frac{\text{Supply Volts}}{\text{Earth Loop Impedance}}$$

Answer the following questions

1. Define: (*a*) Earth electrode.
   (*b*) Earthing lead.
   (*c*) Earth continuity conductor.
2. What are the BASIC REQUIREMENTS for earthing?
3. What is meant by earthing in an electrical installation?
4. What is the meaning of the expression 'Earthed neutral'?

Exercise No. 6.5 — Separate test of earth continuity conductor.

Apparatus

Neutral earth loop tester with provision for earth continuity test.
Low reading ohmmeter.
A lighting circuit wired in metal conduit.

Procedure

1. Switch off main switch and remove fuses.
2. Remove outgoing phase conductor from main switch.
3. Temporarily connect end of outgoing phase conductor to earth so that the switch feed can be traced (e.g. using short length of V.R.I. cable and connector).
4. Proceed to last tumbler switch on the system and use the low-reading ohmmeter to trace the switch feed.
5. Remove the switch feed from the switch and make temporary earth connection as in step number 3.
6. Remove temporary earth connection at main switch and connect outgoing phase conductor to the continuity test terminal on test set.
7. Plug in test set to a separate supply and note reading.
8. Restore all connections to normal.

Answer the following questions

1. When is a separate test of the earth continuity conductor necessary?
2. State the minimum acceptable values of earth continuity impedance stipulated by the I.E.E. Regulations.

EXERCISE No. 6.6 **Earth electrode resistance measurement.**

Apparatus

Demonstration board No. 9.
'Megger' earth tester.

Diagrams

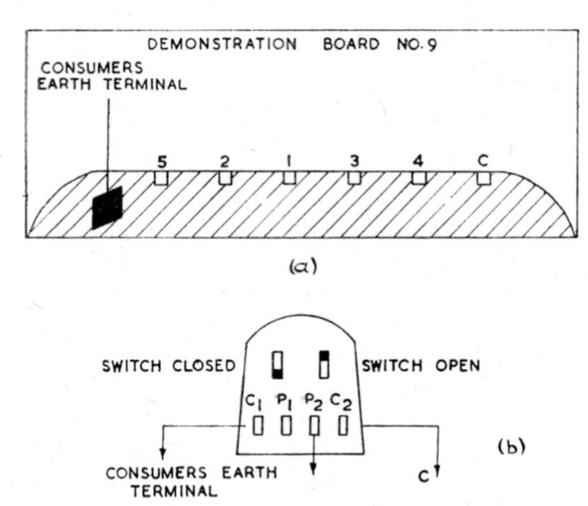

FIG. 6.8 (a–b) Exercise 6.8

Method

1. Connect consumer's earth terminal to terminal C_1 on earth tester.
2. Connect socket C on the board to terminal C_2 on the earth tester.
3. Connect socket 1 on the board to terminal P_2 on the earth tester.
4. Ensure that the switches on the earth tester are in the correct positions as shown in Fig. 6.8(b).
5. Turn the handle on the earth tester at a steady speed and note the reading obtained.
6. Unplug the lead from socket 1 on the board and insert in turn in sockets 2, 3, 4 and 5, taking and noting the reading at each step.

7. Tabulate your results as follows:

Potential Electrode Position	*Ohms*
1	
2	
3	
4	
5	

Average of results 1, 2 and 3 = ... Ohms.

Answer the following questions

1. What is the resistance of the earth electrode tested?
2. Why are the results obtained from positions 4 and 5 unreliable?
3. Why is it desirable to obtain 3 consistent results (as in tests 1, 2 and 3).
4. Describe in your own words how you would measure the earth electrode resistance in practice.

EXERCISE No. 6.7 **To test and inspect a complete installation.**

Apparatus

Model installation in workshop.
Insulation resistance tester.
Low reading ohmmeter.
Voltage indicator.
Neutral/earth loop impedance tester.

Method

(*Note that no details of the tests are given as these are described in previous exercises.*)

1. Ensure that main supply is not connected.
2. Make a careful visual examination of the installation to observe that I.E.E. Regulations have been carried out in respect of: types and sizes of cables used, all covers of switches and joint boxes in position, earth continuity conductors and bonding connected where required, etc.
3. Using the low reading ohmmeter, check the polarity of all switches. Also test for continuity of conductors in ring circuits, and continuity of earth conductors.
4. Using the insulation tester, carry out insulation resistance tests to earth and between conductors, ensuring that all lamps are removed from their holders and that all switches are on.

   If the results of the above tests are satisfactory, then switch off all switches and connect the mains supply.
5. Switch on main switch and test polarity of all lighting switches using the voltage indicator; insert lamp bulbs and check for correct operation of lighting switches.
6. Test socket outlets for effectiveness of earthing using the neutral earth loop tester, this instrument will also indicate correct polarity of the sockets.
7. Check any fixed equipment (e.g electric cookers, water-heaters, etc.) for correct operation.
8. If an earth leakage circuit breaker is fitted, check it for correct operation.
9. Complete a maintenance report of the type below.

Maintenance Report: (*Based on 'Form B' prescribed in I.E.E. Regulations*

I certify that the installation at ....
has been inspected and that:

(*a*) The value of the insulation resistance to earth is .... MΩ
the value of the insulation resistance between conductors is .... MΩ

(*b*) The value of earth-fault loop impedance is ....Ω
or, is earth leakage breaker effective?..........................

(*c*) Earthing is in compliance with the requirements of the regulations, except as stated below.
(*d*) All flexible cords, switches, fuses, plugs and socket outlets are in good serviceable condition, except as stated below.
(*e*) There is no sign of overloading of conductors or accessories except as stated below.
(*f*) There is no evidence of the use of portable appliances in any bathroom except as stated below.
(*g*) There are no obvious defects, and the whole installation appears to be in good serviceable condition, except as stated below.

Signed ....

Date ....

Details of defects and exceptions, if any, referred to above:

Answer the following questions

1. At what intervals should an installation be inspected?
2. Explain in your own words the meaning of 'Insulation Resistance'.

CHAPTER 7

Electric Heating

7.1

One of the three main effects of an electric current is that heat is produced whenever a current is passed through a wire. This heating effect is utilized in many forms of electric heating appliances such as water-heaters, fires, cookers. A heating element consists of resistance wire supported on a 'former' which is capable of withstanding high temperatures and is also a good electrical insulator.

7.2 Water Heating

(*a*) An electric kettle is a good example of an electric water-heater. The type of kettle illustrated by Fig. 7.1, uses an immersion type heating element, consisting of a resistance wire heating coil insulated with magnesia, which is enclosed in a watertight metal tube. The element incorporates a safety device which ejects the connecting adaptor in the event of the element overheating, e.g. if the kettle boils dry or is inadvertently plugged in before being filled.

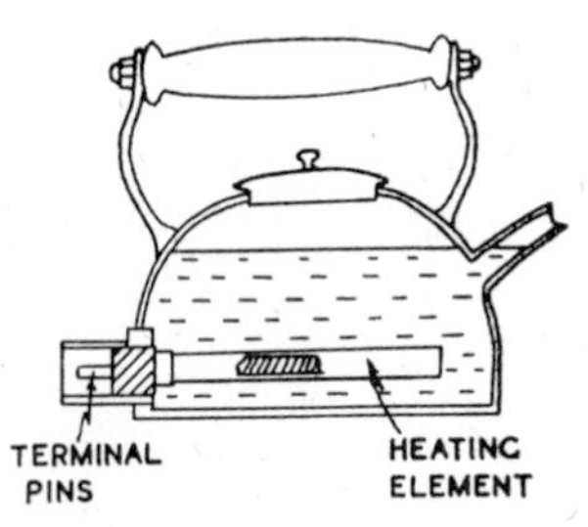

FIG. 7.1 Electric Kettle

(*b*) An immersion heater can be fitted to an existing hot-water tank or cylinder in order to provide a hot-water supply to one or more outlets. Most immersion heaters make use of thermostatic control, the thermostat switch interrupting the supply of electricity to the heating element when the desired temperature is attained, and restoring the supply when the temperature eventually falls. When fitting immersion heaters it must be remembered that hot water rises due to convection and so the water below the level of the heater will not be heated. Heaters fitted to rectangular tanks should in general be mounted horizontally near the bottom of the tank but allowing sufficient space (approximately 3 in.) below the heater to allow for the accumulation of sludge and scale deposits where they will not affect the working of the heater. Heaters fitted to cylindrical storage vessels are mounted from the top and must be long enough to reach well down into the vessel. In many cases it is an advantage to fit the type of heater known as a circulator; this is fitted with a draught tube so that cold water enters at the lower end, is heated and then emerges from the top of the heater. This helps to produce a layer of hot water at the top of the vessel which floats on the colder water in the bottom. Thus when hot water is drawn off from the top, the incoming cold water is directed to the lower part of the vessel, so reducing mixing which would lower the temperature of the hot water. If there are large heat losses from the hot-water storage tank the energy consumed by the immersion heater will be excessive; it is therefore good practice to lag hot-water tanks with heat insulating

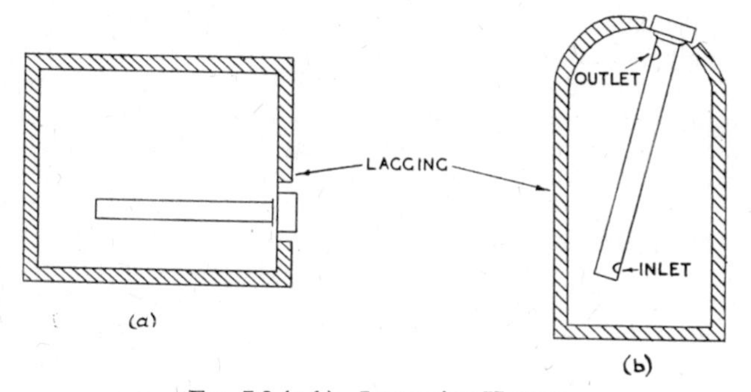

FIG. 7.2 (a–b) Immersion Heaters
(a) Immersion heater in rectangular tank; (b) Circulator in cylindrical tank.

material such as glass-fibre, granulated cork, slagwool, etc. It has been stated that the heat losses from an unlagged 20-gallon storage tank amount to an energy loss of approximately 76 kWh per week, but that this can be reduced to a loss of only 8 kWh per week by lagging with 3-in. thickness of granulated cork, glass-fibre or similar material.

(*c*) 'Non-pressure' type water-heaters are installed to provide a single hot water outlet, e.g. over sinks or hand-basins. The water flow through the heater is regulated by a control valve in the cold water supply pipe. When the cold water is allowed to enter, it displaces the hot water already present in the heater thus causing it to flow from the outlet pipe.

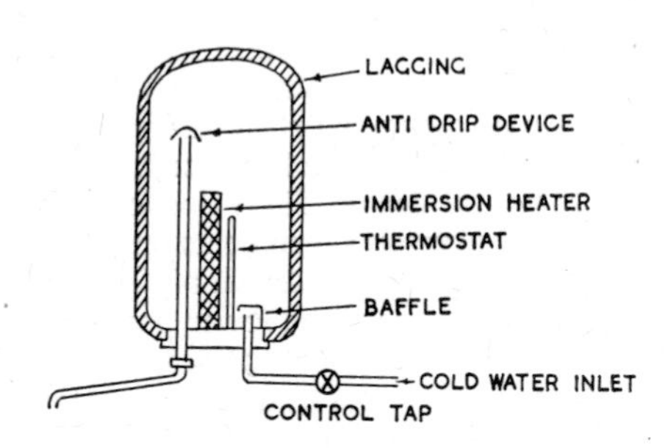

FIG. 7.3 Non-Pressure Type Water Heater

(*d*) 'Pressure type' water-heaters may be used where it is required to supply more than one outlet. They differ from the non-pressure type in that there is no control valve in the cold water inlet pipe. Instead of this a tap is fitted to each outlet. This means that the water in the heater is under pressure provided by the head (or vertical height) of the cold water supply system.

7.3 Space Heaters

Electrical heating may be provided in enclosed spaces such as rooms or workshops, using either 'radiant' heat or 'convected' heat, or both.

(*a*) Radiant heaters utilize heat transmitted in the form of rays which can be directed by means of a reflector to any required area. The heating elements used in this type of heater must operate at red heat in order to produce a good radiant heat output. Fig. 7.5 shows a type of radiant heater often used in workshops.

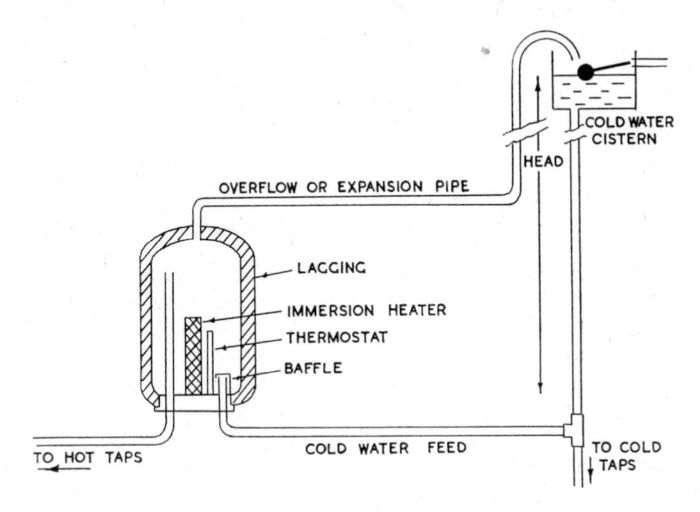

FIG. 7.4 Pressure Type Water Heater

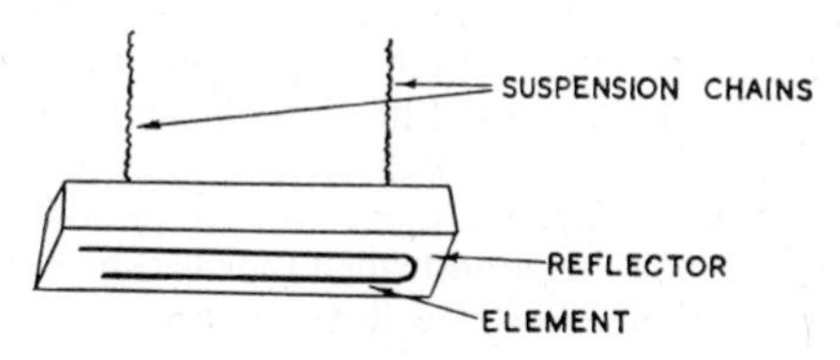

FIG. 7.5 Radiant Heater

(*b*) Convected heat is heat which is conveyed by the movement of warmed particles. The low temperature 'tubular' heater, shown in Fig. 7.6, relies on convection to distribute its heat. Air in the vicinity of the heater is warmed and so rises, its place being taken by colder air; this establishes a convection current of warm air which distributes the heat around the room. This type of heater is often controlled by a thermostat fixed in a suitable position in the room.

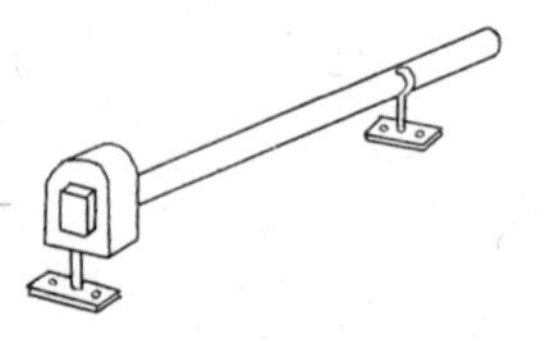

FIG. 7.6 Low Temperature Tubular Heater

(*c*) Convector heaters consist of a heating element fitted in the lower part of a sheet metal case as shown in Fig. 7.7. Cold air enters the bottom of the heater and is warmed by the element; the hot air rises inside the body of the heater and emerges from the louvres at the top.

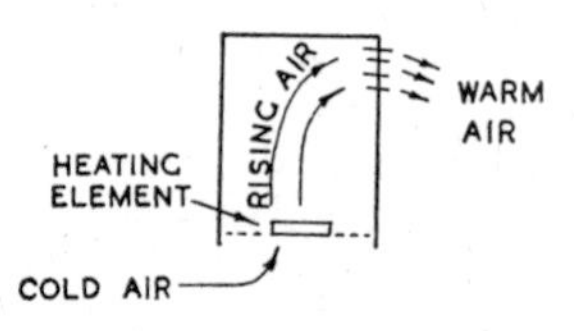

FIG. 7.7 Convector Heater

(*d*) Thermal storage heaters contain a large mass of refractory material which is heated to a high temperature by an electrical heating element. The supply to the heating elements is controlled by a time switch which is set to switch on only during 'off-peak' hours, generally overnight and for a short 'top-up' period during the day. Heat is stored during the period when the heating element is on, and is released at a controlled rate throughout the day. Most supply authorities provide a special 'off-peak' tariff for this type of operation so that this type of heater can provide background heating at an economic price.

7.4 Control of Heating Circuits

(*a*) While many electric heaters are controlled by simple on/off switches, the following are some of the methods which may be used to give better control of the heat produced:

(i) Three-heat switches.
(ii) Simmerstats.
(iii) Thermostats.

Three-heat switch control is used to give a choice of three levels of heating power using a rotary switch in conjunction with two heating elements of equal power rating. A typical circuit arrangement is shown in Fig. 7.8, the internal connections for each position of the switch being indicated by dotted lines.

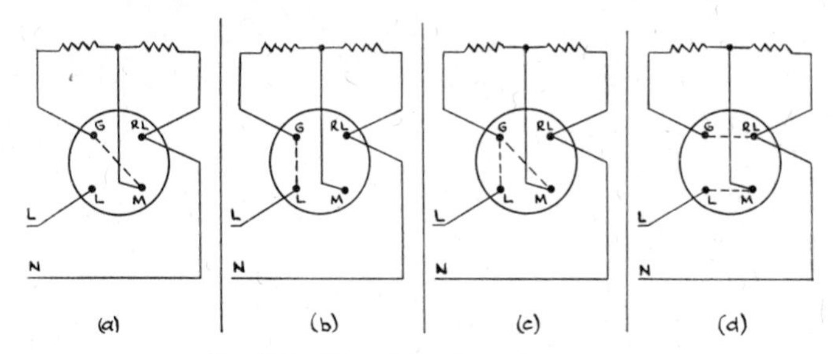

FIG. 7.8 Three-Heat Switch Control

(a) OFF No connection No Power; (b) LOW Elements in series $\frac{1}{4}$ power; (c) MEDIUM one element on $\frac{1}{2}$ power; (d) HIGH Elements in parallel Full power.

(*b*) A simmerstat variable heat switch acts as an 'electrical tap' and provides continuous control of the amount of electrical energy supplied to a heating element. The method of operation of a simmerstat is as follows:

When switched to the 'off' position the control cam depresses the end of the bi-metal strip which turns on the pivot and so holds the contacts open. When the control knob is turned to an intermediate position the control cam allows the end of the bi-metallic strip to rise, so permitting the contacts to close. This completes the circuit to the heating element and also to the heat coil in the simmerstat. The heat produced by the heat coil causes the bi-metallic strip to bend in such a way that the upper and lower leaves move apart, the upper leaf rising freely until it meets the cam. This prevents any further upward movement so the lower leaf moves down, forcing the contacts apart and thus breaking the circuit. The supply to the heating element (and to the heat coil) is interrupted until the bi-metallic strip cools sufficiently to allow the contacts to close again, whereupon the cycle of operation is repeated.

The ratio of the 'on' to 'off' period, which is controlled by the cam setting, determines the average energy consumed by the heating element. For example, if a 2-kW heating element is being controlled and the control knob is set to allow 5 seconds 'on' and 15 seconds 'off' then the average power used is: $2{,}000 \times \frac{5}{15 + 5} = 500$ watts.

In the fully 'on' position of the control knob, the cam allows the bi-metallic strip to rise freely to such an extent that the contacts are never opened, thereby allowing the heating element to consume its full-rated power.

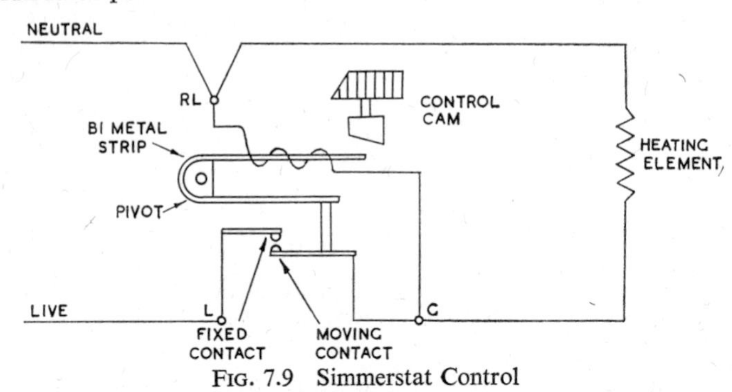

FIG. 7.9 Simmerstat Control

(*c*) Thermostats are used to provide a means of controlling a heating appliance so that a definite temperature is maintained. In practice thermostats maintain the temperature within given limits by switching off the appliance when a desired temperature is attained and switching it on again when the temperature falls by a certain amount. The difference between the switching off and switching on temperatures is called the 'differential' of the thermostat. The principal methods of operating the switch contacts of thermostats are listed below:

METHOD OF OPERATION	APPLICATION
Expansion of metal rod.	Water-heater thermostats.
Expansion of a liquid.	Oven thermostats.
Bending of a bi-metallic strip.	Electric iron and air thermostats.
Expansion of a gas.	Air thermostats.

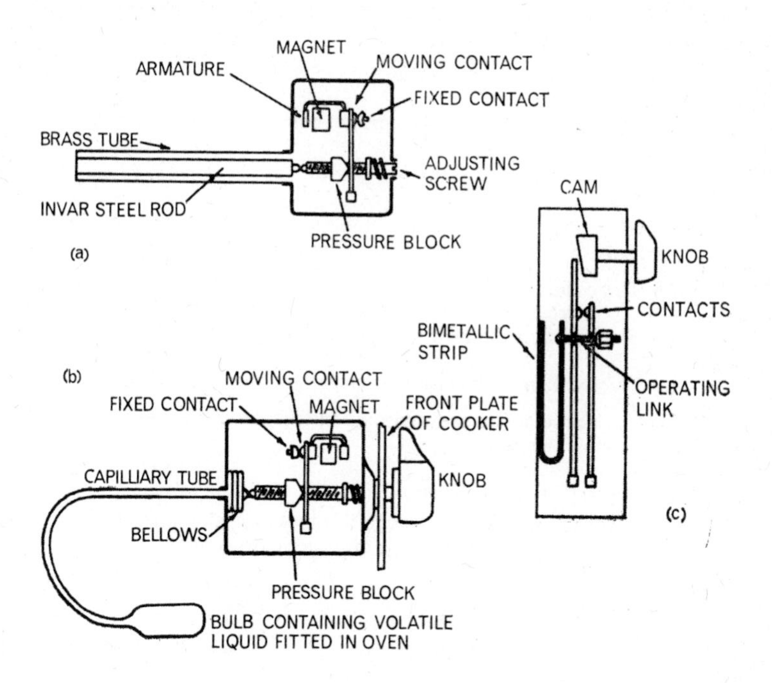

FIG. 7.10 (a–c) Thermostats
(a) Water heater thermostat; (b) Oven thermostat; (c) Air thermostat.

EXERCISES

The exercises which conclude this chapter illustrate some forms of electrical heat control, and the construction of typical water-heaters.

EXERCISE No. 7.1 — To investigate a non-pressure type water heater.

Apparatus

A non-pressure type water-heater; not connected to electricity or water supplies.

Procedure

1. Carefully dismantle the heater and examine its construction paying particular attention to the following points:
   (*a*) The position of the inlet and outlet water pipes and the length of any internal pipes connected to them.
   (*b*) The baffle on the inlet pipe and the anti-drip device on the outlet pipe.
   (*c*) The location of the heating element and thermostat.
2. Make a neat diagram showing the construction of the heater.
3. Reassemble the heater.

Answer the following questions

1. How is the flow of water through this type of heater controlled?
2. For what purposes are this type of heater most suited?

EXERCISE No. 7.2 — To investigate a pressure type water heater.

Apparatus

Small domestic pressure type water heater; not connected to electricity or water supplies.

Procedure

1. Carefully dismantle the heater and examine its construction paying particular attention to the following points:
   (*a*) The position of the water inlet, outlet and overflow connections and the lengths of any internal pipes connected to them.
   (*b*) The baffle on the cold water inlet.
2. Make a neat diagram showing the construction of the heater.
3. Reassemble the heater.

Answer the following questions

1. How does the pressure type water heater differ from the non-pressure type?
2. What is the purpose of the lagging jacket surrounding the heater?

EXERCISE No. 7.3 **Three-heat switch control of heating elements.**

Apparatus

Demonstration board No. 10.
Low reading Ohmmeter.

Diagram

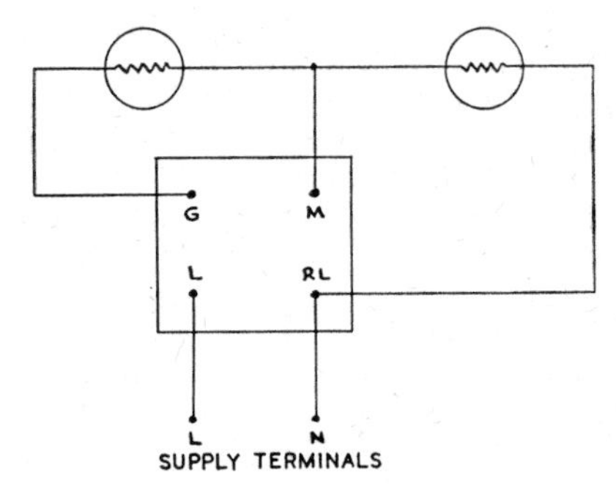

FIG. 7.11 Exercise 7.3

Procedure

1. Connect demonstration board circuit in accordance with Fig. 7.11, noting that 40-watt lamps are used to represent the heating elements.
2. Plug in to main supply and note that:

   (*a*) With switch at 'OFF' both lamps are off.
   (*b*) With switch at 'FULL' both lamps are on brightly.
   (*c*) With switch at 'MED' one lamp is on brightly.
   (*d*) With switch at 'LOW' both lamps glow dimly.

3. Disconnect all wiring and, using the low reading ohmmeter, trace the connection between the switch terminals for each position of the switch.

Answer the following questions

1. Carefully draw four circuit diagrams, one for each position of the switch. The diagrams must show the internal switch connections for each case.
2. Explain the operation of this circuit.
3. If each heating element were rated at 1 kW and the supply voltage was 250V, what total power would be consumed for each position of the switch?

EXERCISE No. 7.4

Simmerstat control of a heating element.

Apparatus

Demonstration board No. 11.

Diagram

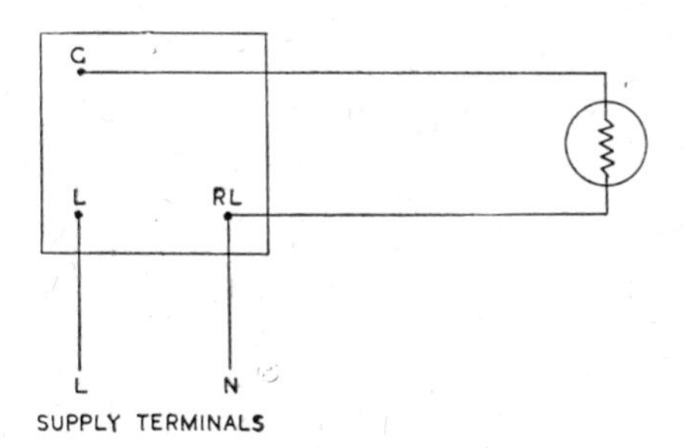

FIG. 7.12 Exercise 7.4

Procedure

1. Connect demonstration board circuit in accordance with Fig. 7.12. Note that a 40-watt lamp is used to represent the heating element.
2. Plug in to main supply and turn control to 'full', note that the lamp remains on continuously.
3. Turn control to 'mid-point' and note that the lamp is switched on and for equal periods of time.
4. Turn control towards zero and note that the lamp is now switched on for a short period and off for a longer period.

Answer the following questions

1. Explain the operation of the simmerstat.
2. What is the difference between a thermostat and a simmerstat?
3. What are the main advantages of the simmerstat as compared with Three-heat Switch Control?

EXERCISE No. 7.5

To connect immersion heater, thermostat and pilot lamp and investigate the operation of the heater.

Apparatus

Ten-gallon water-tank fitted with 4 kW immersion heater. Supply point fitted with kWh meter, main switch, batten holder and fuse.
Thermometer 0–100° C.
Stop-clock.

Diagram

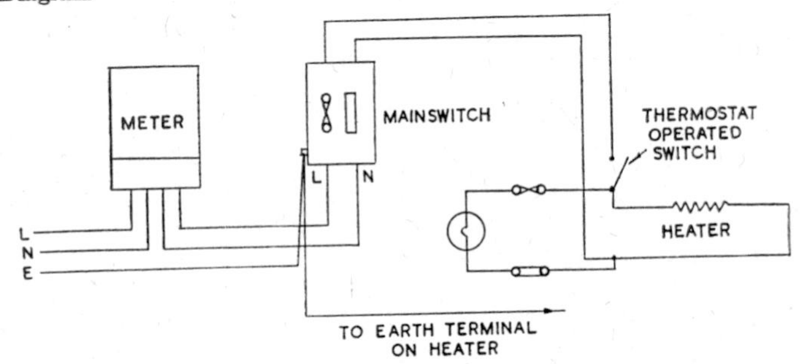

FIG. 7.13 Exercise 7.5

Procedure

1. Connect apparatus as shown in Fig. 7.13.
2. Fill tank with water until it is three-quarters full.
3. Fit thermometer so as to measure temperature of the water in the tank.
4. Note temperature and reading on kWh meter, and switch on supply.
5. Note temperature and meter reading when pilot light goes out, and start stop-clock.
6. Note temperature when pilot light comes on again.
7. When pilot light cuts out again stop the clock and note meter reading.

Results

Initial temperature =
Temperature at which thermostat switches off =
Temperature at which thermostat switches on =
Initial meter reading =
Meter reading when thermostat first switches off =
Meter reading when thermostat next switches off =
Time interval between above two readings =

Answer the following questions

1. Why is the heater fitted low in the tank?
2. What is the average controlled temperature of the water?
3. What is the 'differential' of the thermostat used?
4. How much electrical energy was used initially to heat the water?
5. How much electrical energy was needed to make up the heat losses between thermostat operations?
6. Calculate the average electrical power used in overcoming heat losses.
7. How can the heat losses be reduced to a minimum?

EXERCISE No. 7.6 — To investigate the operation of an oven thermostat.

Apparatus

Electric cooker fitted with oven thermostat and thermometer.

Procedure

1. Set thermostat to 500 mark and switch on.
2. Note oven temperature at five-minute intervals for thirty minutes.
3. Plot a graph of temperature against a base of time.

Answer the following questions

1. How long did the oven take to attain its working temperature?
2. Between what limits did the thermostat hold the temperature?
3. What was the average temperature?
4. Explain with the aid of a neat sketch the action of the oven thermostat.

CHAPTER 8

Lamps and Lamp Circuits

8.1 Incandescent Lamps

Incandescent lamps have a filament which is heated to white heat by the passage of an electric current. The filaments of modern lamps are normally made of tungsten since this material has a very high melting point (3,400° C.) and can be manufactured in the form of a suitably thin wire. The bulbs of smaller lamps are evacuated to prevent oxidization of the filament but, in many lamps, an inert gas such as ARGON is introduced. This enables the filament to operate at a higher temperature without undue deterioration due to the evaporation which tends to take place in a vacuum. The filaments of many lamps are constructed using the 'coiled-coil' principle, the main advantages being:

(*a*) The filament has a more compact formation.

(*b*) Heat losses due to convection currents in the gas are reduced, thus giving a higher efficiency.

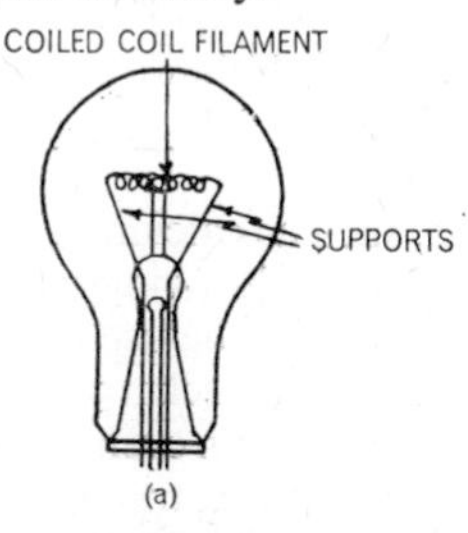

FIG. 8.1 (a–b) Incandescent Filament Lamp
(a) Lamp; (b) Coiled coil filament.

8.2 Discharge Lamps

(*a*) When an electric current is passed through certain gases visible light is produced. The gas is normally contained in a long tube fitted with an 'electrode' at each end. A fairly high voltage is required to maintain the discharge although the current is small. For most discharge lamps the 'striking' voltage required is higher than the 'running' voltage and so some means of limiting the running current is required. Several examples of the methods used are illustrated in the circuits at the end of this chapter. The light output of simple discharge lamps is, in general, very low and these lamps are commonly used for advertising signs, small neon tubes being frequently used as indicator lamps. The colour of the light emitted depends upon the type of gas used, the colours obtained from some of the gases and vapours commonly employed are listed in the table below:

Colours of Discharge Lamps

Gas or Vapour	*Colour of Light*
Neon.	Red.
Hydrogen.	Pink.
Helium.	Ivory.
Nitrogen.	Buff.
Mercury (low pressure).	Blue together with strong ultra-violet emission.
Mercury (high pressure).	Bluish white with less ultra-violet emission.
Sodium.	Yellow.

(*b*) Certain materials, such as calcium halophosphate, emit visible light whenever they absorb ultra-violet light. This phenomena is known as fluorescence and may be used to produce a very efficient type of lamp. If the tube of a discharge lamp containing mercury vapour is coated internally with an even layer of fluorescent material a considerable proportion of the ultra-violet light caused by the discharge is converted into useful visible light.

(*c*) In some types of discharge lamp the electrodes are not heated. These types are therefore known as 'cold cathode' lamps, an example of this being the ordinary neon tube. In other types of discharge lamp the electrodes are heated, as this reduces the voltage required to strike and maintain the discharge. Lamps using heated electrodes are known as 'hot cathode' lamps a typical example being the ordinary fluorescent lamp. The 'hot cathodes' are usually in the form of a short filament which may be heated either by passing a heating current through it or by the discharge current itself.

8.3 Fluorescent Lamp Circuits

(*a*) Fig. 8.2 shows the basic 'switch start' circuit which operates as follows:

When the supply is switched on with the starter switch, S, closed a current flows through the inductor, L, and through the lamp electrodes, E. This initial current heats the lamp electrodes in readiness for striking the lamp. The starting switch is now opened making a sudden interruption in the current flowing through the inductor and so causing a high voltage to be momentarily induced. This voltage starts a discharge between the two lamp electrodes and the current rapidly rises to a value determined mainly by the inductance of the inductor. The starter switch is left open while the lamp is alight, the electrodes maintaining their operating temperature as long as they continue to pass the discharge current. In practice it is desirable that the starter switch should operate automatically, switching on when the supply is first switched on then switching off to strike the lamp and remaining off all the time that the lamp is alight. There are two types of starter switch in general use, the glow type and the thermal type. Due to the inductor the lamp current lags the supply voltage (at approximately 0·5 p.f.). A capacitor, C, is usually connected between the lamp terminals to improve the overall power to an acceptable value.

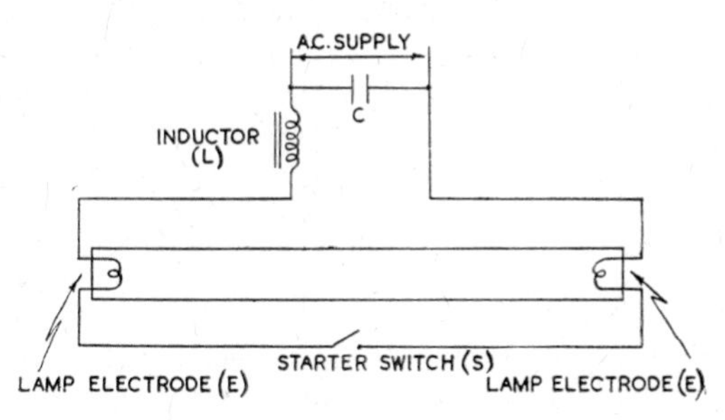

FIG. 8.2 Basic Fluorescent Lamp Circuit

(*b*) The glow type starter switch consists of a small bulb filled with helium and containing two contacts, one of which is mounted on a bi-metal strip. The contacts are normally open so that when the mains supply is first switched on full mains voltage is applied to the starter contacts. This causes a glow discharge which warms the bi-metal strip making it bend, so closing the starter contacts. The closing of the starter contacts allows full heating current to pass through the lamp electrodes and also extinguishes the glow discharge. After a short time the bi-metal strip cools sufficiently to open the circuit thus striking the lamp. As long as the lamp remains alight the voltage applied to the starter is insufficient to initiate a glow discharge and so the starter contacts remain open until the next starting operation. A small capacitor is often connected in parallel with the starter switch contacts to suppress radio interference.

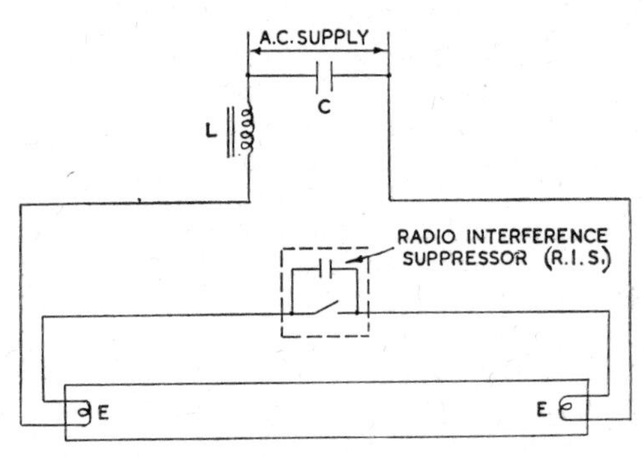

FIG. 8.3 Glow Starter Circuit

(*c*) The thermal type starter switch has two contacts mounted on bi-metal strips, a small heating coil being fitted very close to the bi-metal strips but not in electrical contact with them. The contacts are normally closed so that when the main supply is first switched on full heating current passes through the lamp electrodes as before. The current also flows through the starter heater and so warms the bi-metal strips. After a short time the bi-metal strips warm sufficiently to bend and open the contacts thus striking the lamp. As long as the lamp remains alight current flows through the starter heater keeping the contacts apart.

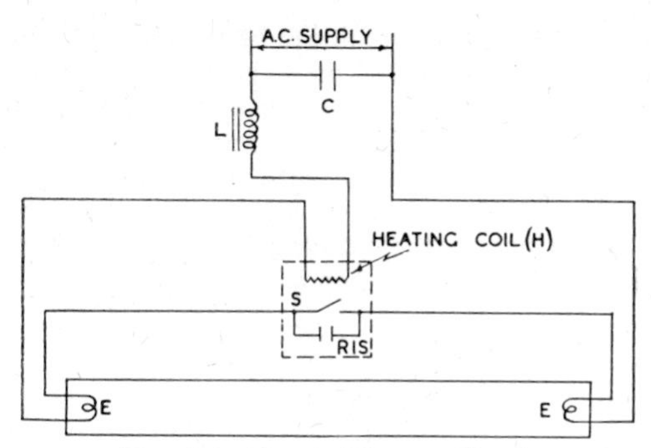

FIG. 8.4 Thermal Starter Circuit

(*d*) Modern types of starter are enclosed in a metal canister having four connecting pins which plug into a standard size socket. There are two large connecting pins which are always connected to the starter switch contacts. The two smaller pins are joined internally for a glow type starter but are connected to the heater coil in a thermal type starter. By using a standard starter socket wired as shown in Fig. 8.5, then either glow or thermal type starting switches may be used.

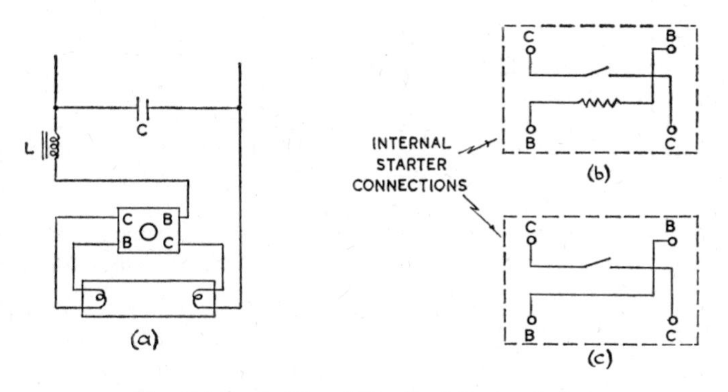

FIG. 8.5 (a–c) Use of Standard Starter Socket
(a) Socket connections; (b) Thermal starter; (c) Glow starter.

(*e*) An instant start lamp circuit can be employed so that the lamp lights almost instantaneously when the circuit switch is closed. There are no automatic starter switches employed in the above circuit, the electrodes being heated by the action of the auto transformer. A narrow earthed metallic strip, which is fixed to the outside of the tube, is necessary to ensure consistent starting, the effect of this strip being to increase the voltage gradient in the mercury vapour near the electrodes when the lamp is first switched on.

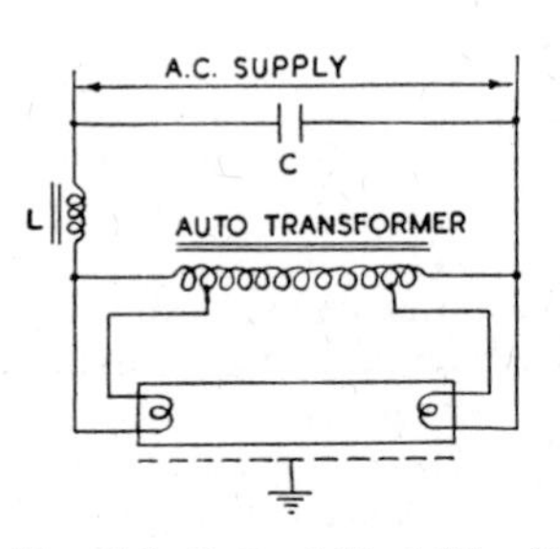

FIG. 8.6 Instant Start Circuit

8.4 Stroboscopic Effects

(*a*) A disadvantage of fluorescent lamps is that as the alternating discharge current passes through zero twice every cycle the light produced tends to flicker at twice mains frequency. Although this effect is not noticeable to the eye, machinery rotating at certain speeds may appear to be stationary or moving more slowly than it really is. This is known as the stroboscopic effect and is obviously a cause of danger in situations such as workshops where rotating machinery is in use.

(*b*) If a three-phase supply is available the stroboscopic effect can be minimized by connecting lamps to alternate phases as shown in Fig. 8.8. As the lamps in this circuit attain their maximum and minimum values of light output in sequence the overall illumination is kept practically constant thereby keeping the stroboscopic effect to a minimum.

(*c*) If only a single phase supply is available then the 'lead lag' circuit shown in Fig. 8.9(a) may be used. In this circuit lamp, A, is

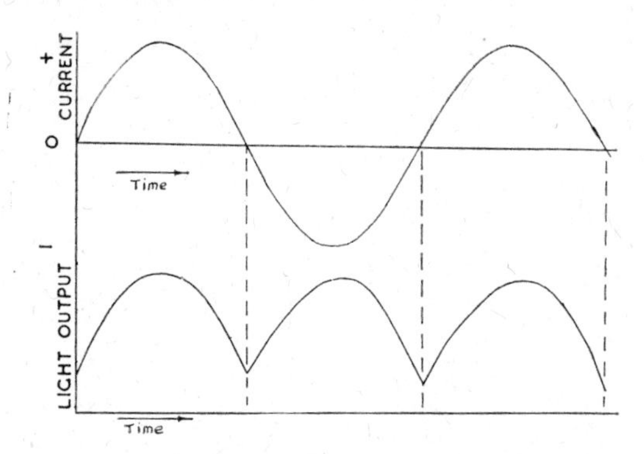

FIG. 8.7 Variation of Current and Light Output of a Fluorescent Lamp Operating from a.c. mains

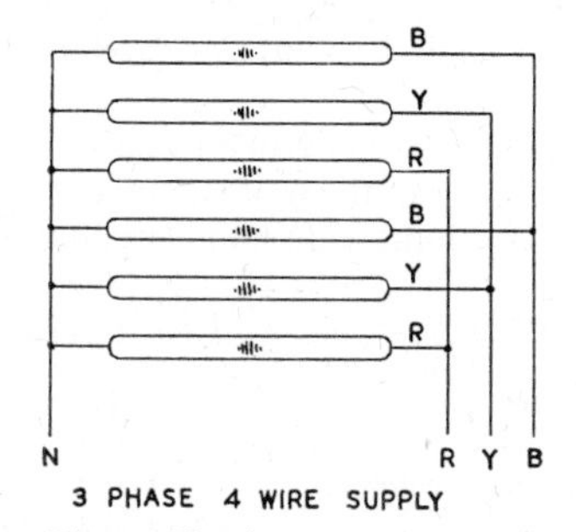

FIG. 8.8 Three Phase Circuit to Reduce Stroboscopic Effect

supplied via an inductor and so has a lagging current. Both an inductor and a capacitor are connected in series with lamp B. The inductor is required to supply the initial starting surge but, when the lamp is alight, the effect of the capacitor predominates so that the lamp takes a leading current. It follows that when one lamp is producing its minimum light output the other is producing its maximum and so, by using this circuit, the stroboscopic effect is greatly reduced.

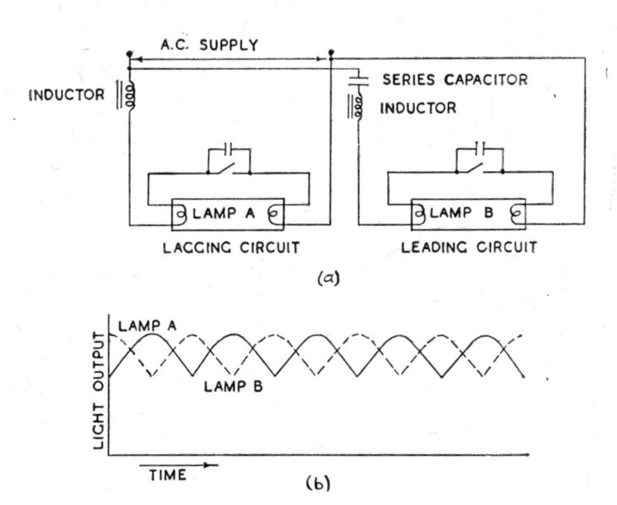

FIG. 8.9 (a–b) Lead Lag Circuit to Reduce Stroboscopic Effect (a) Circuit diagram; (b) Waveforms.

8.5 Fault-finding in Fluorescent Lamp Circuits

The table below summarizes the more common types of fault which may be encountered in fluorescent lamp circuits.

Faults in Fluorescent Lamp Circuits

Symptom	*Possible Cause*
Fuse blows when lamp is switched on.	A faulty p.f. capacitor or short circuit in wiring.
Lamp appears completely dead when first switched on.	No supply, broken tube electrode, faulty starter switch.
Lamp makes repeated efforts to start.	Worn out lamp, low voltage, faulty starter switch.
Lamp does not light but both electrodes glow continuously.	Faulty starter or faulty R.I.S. capacitor.

8.6 High-Pressure Mercury Vapour Lamp (H.P.M.V.)

(*a*) This type of lamp is often used for street lighting and flood lighting purposes because it provides an efficient and compact source with a high light output. It is not suitable for general use as the light it provides is deficient in red which has the effect of distorting the colours of objects as seen by the light from the lamp.

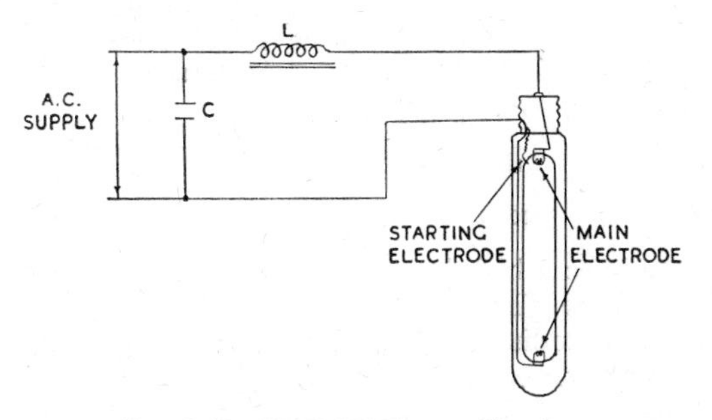

FIG. 8.10 H.P.M.V. Lamp Circuit

(*b*) The starting of this type of lamp is dependent on the fact that when the lamp is cold the pressure in the tube is fairly low so that, when the supply is switched on, a small discharge occurs between the starting electrode and the adjacent main electrode. This discharge spreads along the tube until it reaches the main electrode at the other end of the tube, and the main discharge starts. The tube now heats up so increasing the internal pressure which gives rise to a big increase in light output. After the lamp is switched off it cannot be restarted immediately because the internal pressure is too high until the lamp has cooled down.

8.7 Sodium Lamp

(*a*) This type of lamp is also used for street lighting and similar purposes. Although the sodium lamp is slightly more efficient than the high-pressure mercury vapour lamp the 'colour rendering' of objects seen by its light is even worse, as the light output consists only of yellow light. The yellow light of the sodium lamp is often used to advantage in flood lighting schemes, giving a warmer effect than that obtained using the cold blue light from a high pressure mercury vapour lamp.

(*b*) To start this lamp an auto-transformer is used to provide a high voltage between the lamp electrodes, the transformer being

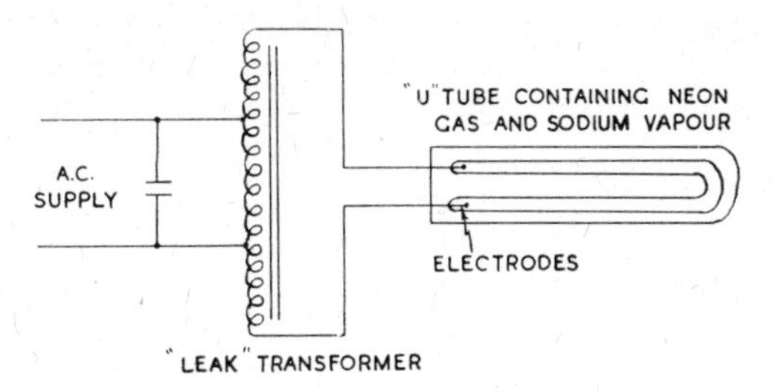

FIG. 8.11 Sodium Lamp Circuit

designed to have a high leakage reactance so that no separate inductor is required. The 'U' tube in the lamp bulb contains a mixture of neon gas and sodium vapour and, when the lamp is cold, some of the sodium vapour condenses to form globules of metallic sodium on the glass of the 'U' tube. When the lamp is first switched on, the high voltage applied to the lamp by the transformer causes a discharge through the neon gas. As the lamp warms up, the globules of metallic sodium vaporize, causing the pressure of the sodium vapour in the lamp to rise and the discharge transfers to the sodium vapour in preference to the neon gas, the light output from the lamp increasing as it warms up, until the full internal vapour pressure is attained. This type of lamp has the advantage that it will strike immediately it is switched on whether hot or cold. When the lamp is switched off, some of the sodium vapour will condense as the lamp cools, and lamps are designed so that the vapour does not condense on to the lamp electrodes, this makes it important that lamps should be mounted only in the attitude for which they were designed, i.e. a lamp designed for 'cap up' mounting should not be used in a fitting where the cap of the lamp is at the bottom.

EXERCISES

The exercises which conclude this chapter are designed to illustrate some common types of discharge lamp circuit.

EXERCISE No. 8.1 — To investigate a switch-start fluorescent lamp circuit.

Apparatus

Demonstration board No. 12.
Ammeter range 0–3A a.c.

Diagram

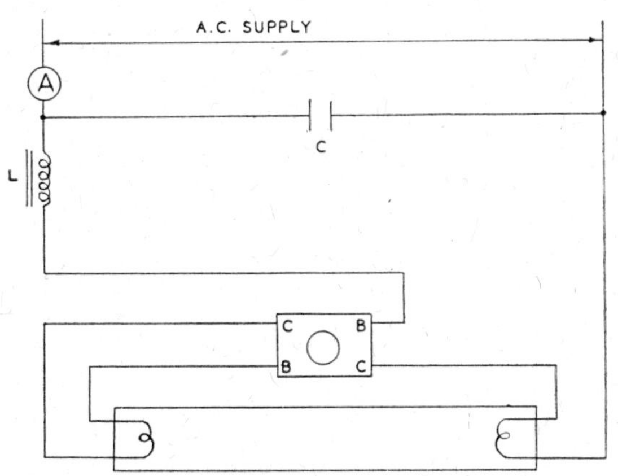

FIG. 8.12 Exercise 8.1

Procedure

1. Connect the circuit as shown in Fig. 8.12.
2. Switch on and note the ammeter reading.
3. Switch off, disconnect p.f. capacitor, switch on and again note ammeter reading.

Answer the following questions

1. What is the purpose of the p.f. capacitor?
2. Explain briefly the operation of a switch-start fluorescent lamp circuit.
3. What are the requirements of the I.E.E. Regulations regarding the rating of final sub-circuits exclusively supplying inductor-operated discharge lamps?

EXERCISE No. 8.2 **To investigate the lead lag fluorescent lamp circuit.**

Apparatus

Demonstration board No. 13.

Diagram

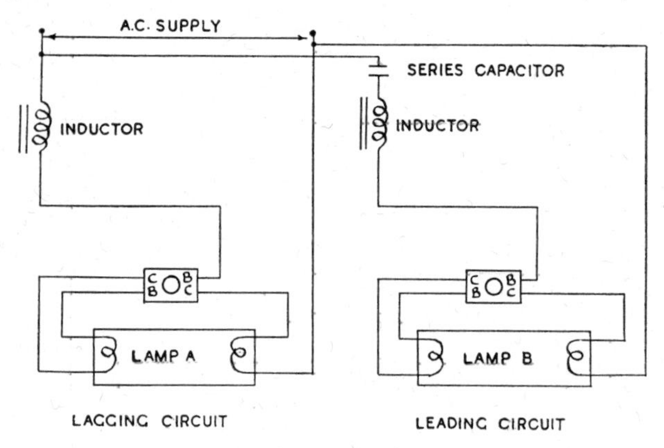

FIG. 8.13 Exercise 8.2

Procedure

1. Connect the circuit as shown in Fig. 8.13.
2. Switch on the supply and note that the lamps operate correctly.

Answer the following questions

1. (*a*) What is meant by the term 'stroboscopic effect'?
   (*b*) Describe two methods of reducing the stroboscopic effect.
2. What are the requirements of the I.E.E. Regulations regarding the screening of the live parts of a discharge lamp installation?

EXERCISE No. 8.3

To investigate an instant start fluorescent lamp circuit.

Apparatus

Demonstration board No. 14.

Diagram

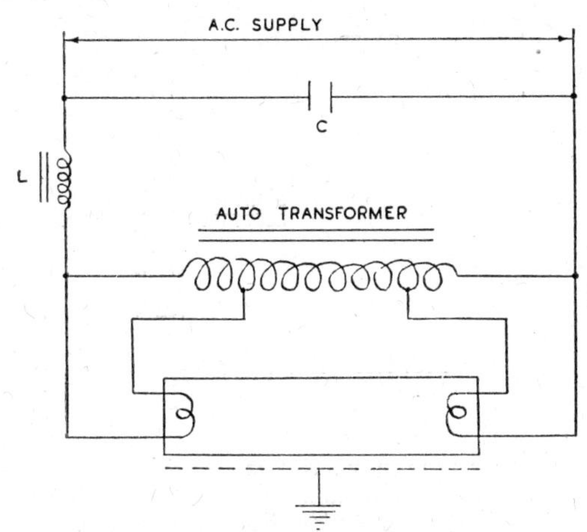

FIG. 8.14 Exercise 8.3

Procedure

1. Connect the circuit as shown in Fig. 8.14.
2. Switch on and note that the lamp lights almost instantaneously.

Answer the following questions

1. Explain briefly the operation of the instant start circuit.
2. Why should tubes used in instant start circuits be fitted with an earthed metallic strip
3. What are the requirements of the I.E.E. Regulations regarding the current rating of switches used to control discharge lamp circuits?

Exercise No. 8.4 — To investigate the high-pressure mercury vapour lamp circuit.

Apparatus

Demonstration board No. 15.

Diagram

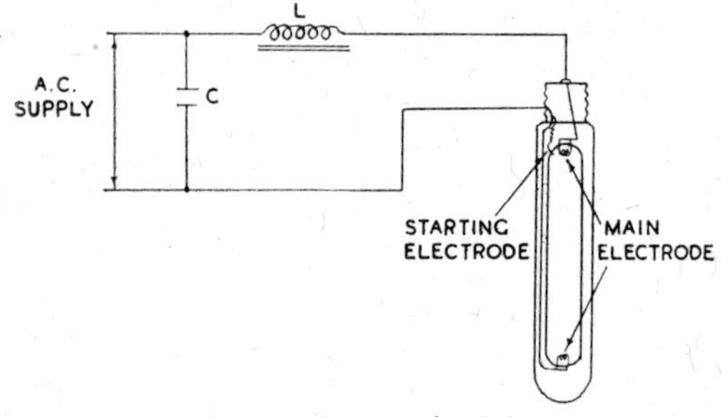

Fig. 8.15 Exercise 8.4

Procedure

1. Connect the circuit as shown in Fig. 8.15.
2. Switch on and leave for approximately ten minutes; switch off and immediately switch on again noting that the lamp will not strike for some time.

Answer the following questions

1. Explain how the discharge is initiated in this type of lamp. What is the reason for the lamp refusing to strike when hot?
2. For what purposes are this type of lamp best suited?

EXERCISE No. 8.5 **To investigate the sodium lamp circuit.**

Apparatus

Demonstration board No. 16.

Diagram

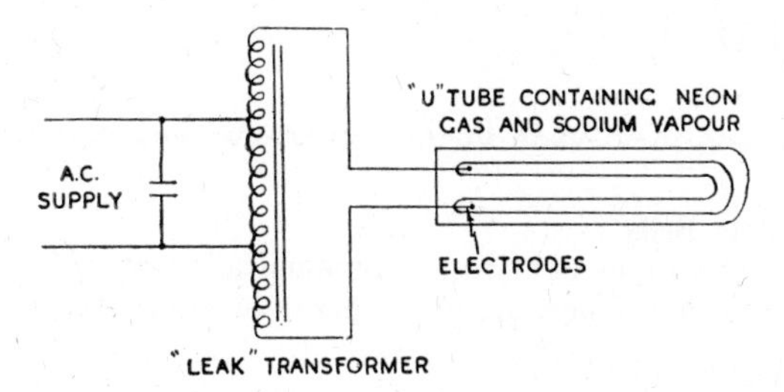

FIG. 8.16 Exercise 8.5

Procedure

1. Connect the circuit as shown in Fig. 8.16.
2. Switch on and observe the change in colour and power of the light as the lamp warms up.
3. After the lamp has been in operation for about ten minutes, switch off. Switch on again and note that the lamp strikes immediately.

Answer the following questions

1. Why is the use of this type of lamp limited to street lighting and similar applications?
2. What are the requirements of the I.E.E. Regulations regarding the design of discharge lamp equipment which under steady running conditions operates at an r.m.s. voltage exceeding 300V but not exceeding 650V on open circuit?

CHAPTER 9

Bell, Alarm and Communication Circuits

9.1 Electric Bells

(*a*) The simplest type of alarm or communicating device is provided by an electric bell. There are four types of bell in common use:

(i) Single-stroke bell.
(ii) Trembler bell.
(iii) Continuous ringing bell.
(iv) Polarized bell.

Fig. 9.1 shows a single-stroke bell. When a current flows through the coils the armature is attracted so ringing the bell. The bell gives one chime every time the push is operated. This type of bell is employed when it is necessary to connect bells in series.

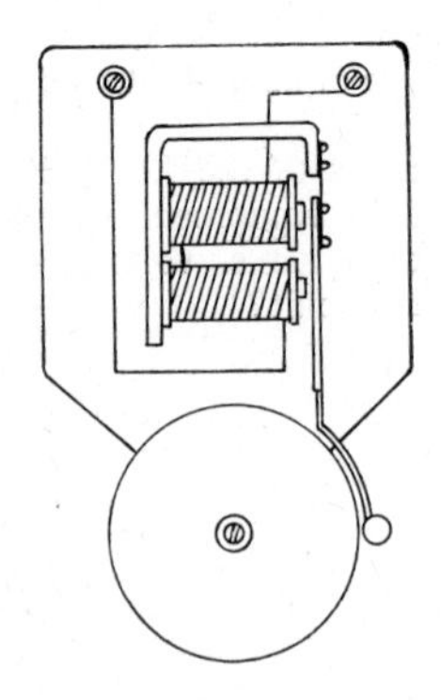

FIG. 9.1 Single-stroke Bell

(*b*) Fig. 9.2 shows a trembler bell. When a current flows through the coils the armature is attracted but the movement of the armature opens the make and break contacts so interrupting the current. The armature is released and moves away from the magnet poles under the influence of the spring. This closes the make and break contacts once more and the action is then repeated. Therefore the armature continues to 'tremble', ringing the bell until the current is switched off. The mechanism of a buzzer is similar to that of a trembler bell with striker and gong omitted.

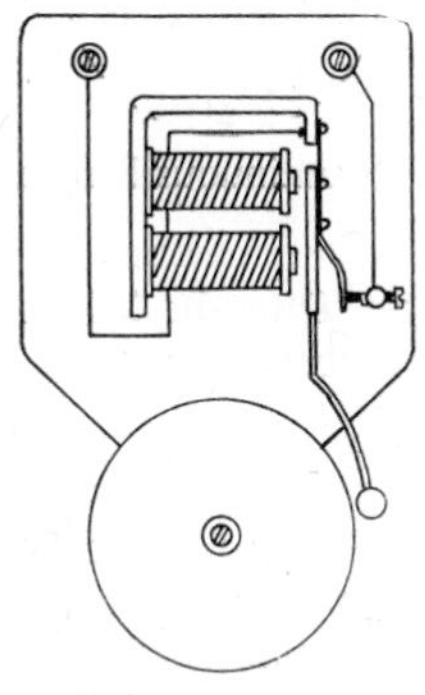

FIG. 9.2 Trembler Bell

(*c*) Fig. 9.3 shows a continuous ringing bell. When the current first flows through the coils the armature is attracted and so releases the catch allowing the contact arm to move under the influence of the auxiliary spring. This closes the contact thus completing the bell circuit independently of the bell-push, and so the bell continues to ring even though the bell-push may be released. The bell will continue to ring until the reset cord is pulled.

(*d*) Fig. 9.4 shows a polarized bell. This type of bell is operated by an alternating current such as may be obtained from the magnetos used in certain types of telephone circuit. When the current in the coils has one polarity, pole A of the magnet is strengthened and pole B weakened; this causes the soft iron armature to pivot so that the striker rings bell 1. When the current reverses, pole A is weakened and pole B strengthened, so the armature moves back

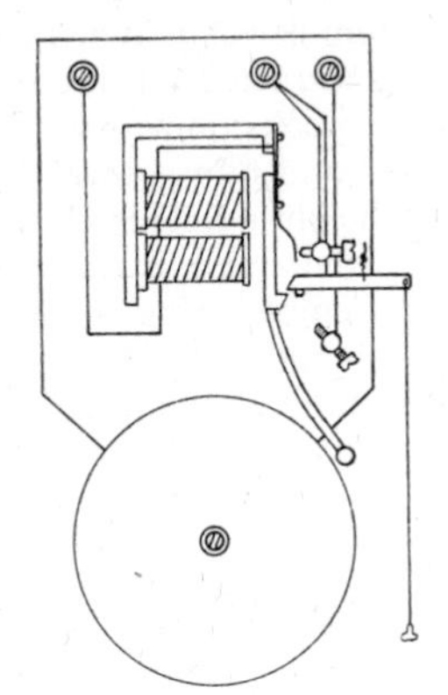

FIG. 9.3 Continuous Ringing Bell

causing bell 2 to be struck. Thus the bell produces two chimes for each cycle of alternating current.

9.2 Simple Bell Circuits

(*a*) The simplest possible bell circuit consists of the bell, battery and bell-push connected in series as shown in Fig. 9.5.

(*b*) Bell circuits may be supplied using a bell transformer instead of a battery as shown in Fig. 9.6. Class A bell transformers provide a choice of three secondary voltages, 4, 8 or 12 volts. Class B bell transformers provide a single secondary voltage of 6 volts. The

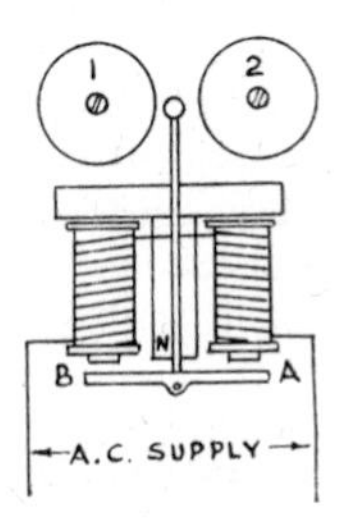

FIG. 9.4 Polarized Bell

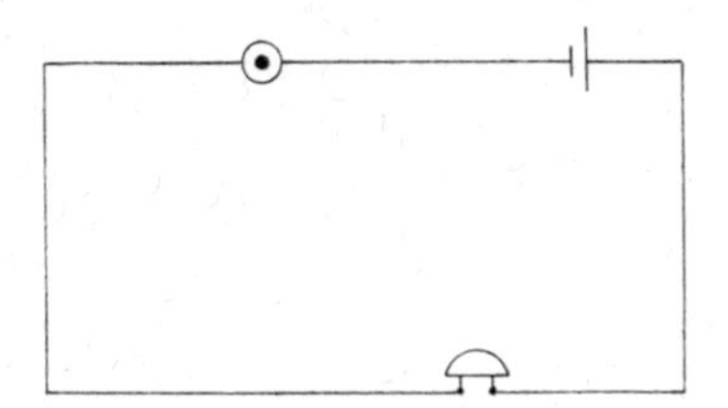

FIG. 9.5 Simple Bell Circuit

voltage used depends upon the type of bell employed and the length of wiring run in the bell circuit. In order to comply with the I.E.E. Regulations the following points should be observed:

(i) The transformer must be double-wound.
(ii) The core of the transformer (metal case if used) and one point of the secondary winding must be earthed.
(iii) The transformer should be connected to a separate sub-circuit and in this event does not require a separate control switch.
(iv) The cables used to supply the transformer must be of a grade suitable for the supply voltage in use.
(v) The secondary wiring need only be insulated for extra low voltage provided that it is completely segregated from power and lighting cables. If it is necessary to run the bell circuit wiring in the same conduit or duct as power or lighting cables, then the bell circuit must employ cables of a grade suitable for the highest voltage present in the power and lighting cables.

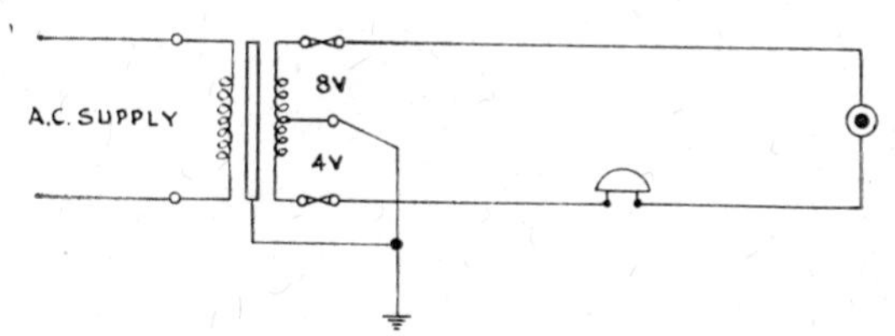

FIG. 9.6 Bell Circuit Using Bell Transformer

9.3 Indicators

(*a*) A bell circuit is often arranged so that the bell may be operated from any one of a number of bell-pushes. An indicator board can then be used to show which bell-push has been operated. A typical circuit incorporating an indicator board is shown in Fig. 9.7. In this circuit, when any particular push is pressed current flows through the indicator element concerned, to the bell.

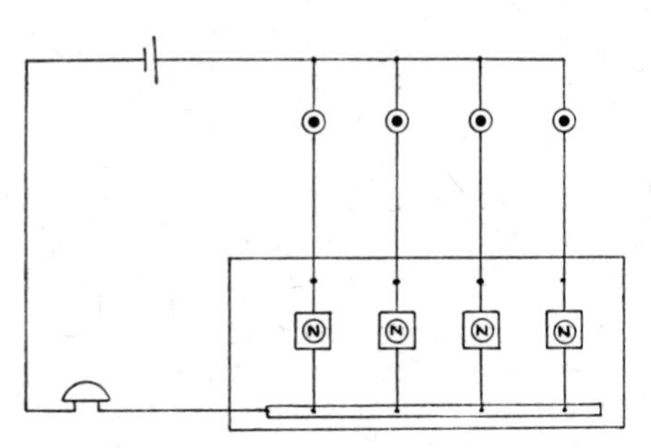

FIG. 9.7 Bell and Indicator Circuit

(*b*) Fig. 9.8 shows a pendulum-type indicator. When the current flows through the coil the armature is attracted. When the current ceases the armature is released and the flag will then swing backwards and forwards for some time.

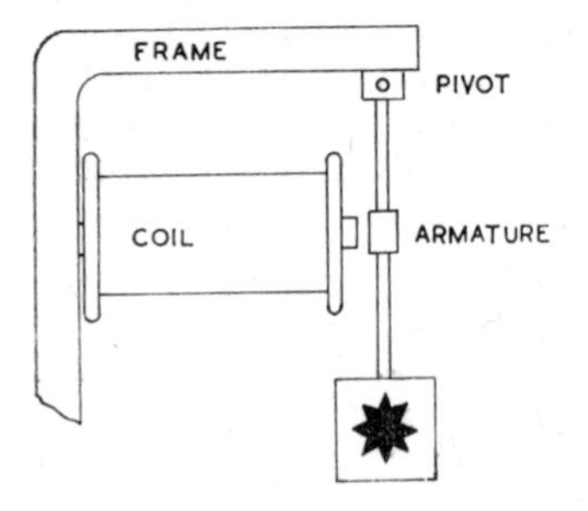

FIG. 9.8 Pendulum Type Indicator

(*c*) Fig. 9.9 shows a mechanical replacement indicator. When the current flows through the coil the armature is attracted. This releases the catch and the flag drops. When the current ceases the armature returns to its normal position owing to the action of the spring, but the flag remains down until it is raised by operating the replacement button.

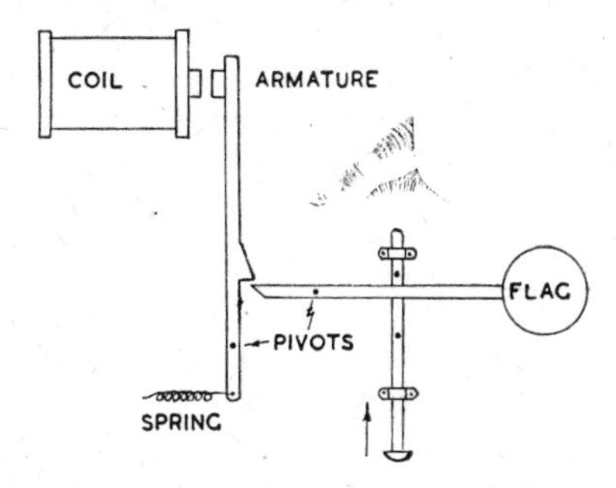

FIG. 9.9 Mechanical Replacement Indicator

9.4 Relays

(*a*) A relay is an electro-magnetically operated switch. A typical relay is illustrated by Fig. 9.10; when current flows through the coil the armature is attracted and operates the contacts. The relay shown has 'normally open' contacts which are closed by the action of this relay. Other types of contact may be fitted, including 'normally closed' which are opened by the action of the relay, and 'changeover' types. Several sets of contacts may be operated by one relay if required.

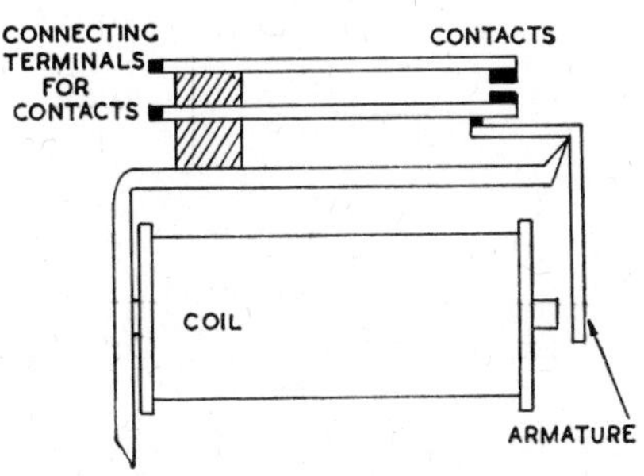

FIG. 9.10 Relay

(*b*) Relays can be designed to operate with a very small current in the coil, and so a small current can be made to switch a larger current on or off. For example, a relay may be used in a circuit where the bell is a long distance from the bell-push, as shown in Fig. 9.11. When the push is pressed the current in the relay coil is sufficient to operate the relay even though the resistance of the wires connecting the bell-push is high because of their length. Operation of the relay causes the contacts to close so switching on the local bell circuit. If it is required to use only one battery the circuit may be modified as shown in Fig. 9.12.

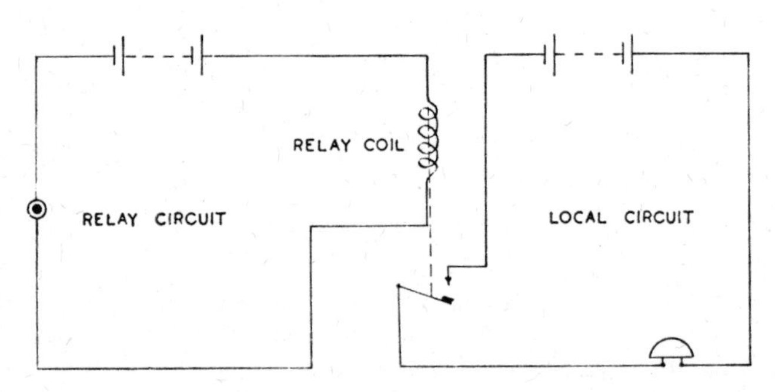

FIG. 9.11 Relay Circuit

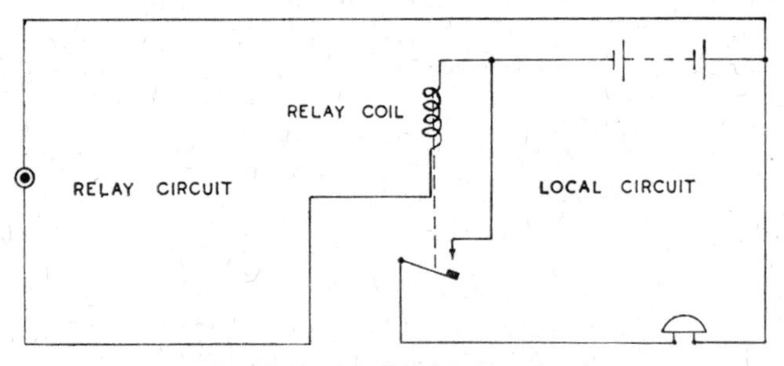

FIG. 9.12 Modified Relay Circuit

(*c*) A relay may be used in conjunction with an ordinary trembler bell to provide a continuous ringing action as illustrated by the circuit shown in Fig. 9.13. The relay is provided with two normally open contacts A and B. When the bell-push is pressed the relay is energized thus closing the contacts; contact A switches on the bell as in the previous circuit. Contact B now completes the circuit to the relay coil independently of the bell-push, so that the relay coil remains energized after the bell-push is released and the bell continues to ring. The circuit can be reset electrically by using a normally closed push-button switch, C. When the reset button is pressed the circuit to the relay coil is interrupted and the contacts A and B return to their normally open condition.

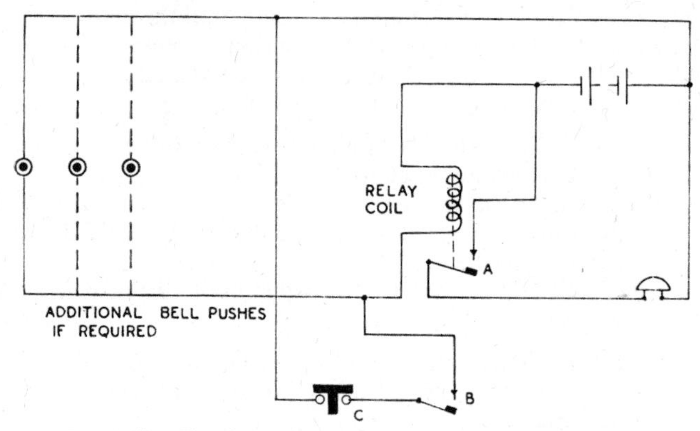

FIG. 9.13 Circuit to Provide Continuous Ringing Action

(*d*) The continuous ringing circuit, shown in Fig. 9.13, can be used as an alarm system by using several bell-pushes wired in parallel. This is known as an 'open circuit' type alarm system as no current flows until one or other of the bell-pushes is operated. This type of circuit could be used to provide a simple fire alarm as operating any push would set the bell ringing continuously. The system has the advantage that no current is used until the alarm is operated. The big disadvantage is that any break in the wiring or a faulty bell-push does not become apparent until the system is operated.

(*e*) A closed circuit alarm system employs a relay which is normally energized by the current flowing around a closed circuit. When the energizing current is interrupted by the action of an alarm contact the relay armature is released so closing the contacts which operate the alarm bell. The operation of a simple closed circuit system is illustrated by Fig. 9.14. This circuit employs a relay with one 'normally closed' contact A and one 'normally open' contact

B. In normal use a current flows through the alarm contacts, contact B of the relay, and the relay coil, thus maintaining the relay in the energized condition so that contact A is open and contact B closed, as shown in the diagram. If any one of the alarm contacts is operated the current through the relay coil is interrupted so releasing the relay armature. This closes contact A and rings the bell, while contact B opens, breaking the relay coil circuit. As contact B is open the relay cannot become energized again even if the alarm contact closes, so the bell continues to ring until the alarm contact is returned to normal and the reset button is pressed. A master switch is provided so that the circuit can be switched off when not required. The advantages of this type of alarm are:

(i) Any failure in the alarm circuit wiring will de-energize the relay and operate the alarm.
(ii) Every time the master switch is closed the bell circuit is automatically checked as the bell should ring until the reset button is pressed.
(iii) The bell cannot be stopped ringing until all the alarm contacts are closed.

This type of circuit can be used to provide an effective burglar alarm system, using concealed alarm contacts fitted to windows, doors, safes, etc.

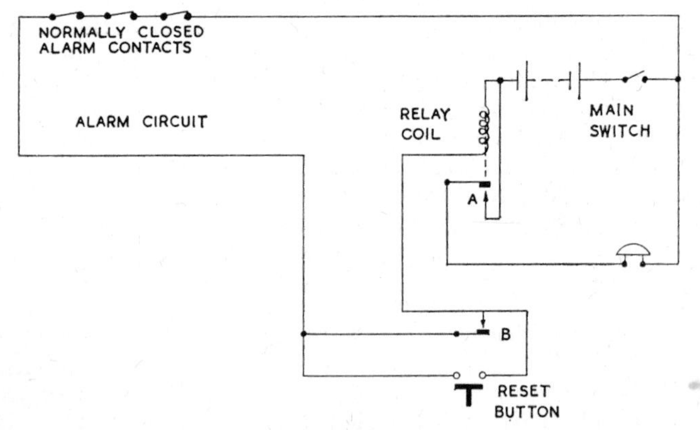

FIG. 9.14 Closed Circuit Alarm System

9.5 Telephone Circuits

(*a*) The essential parts of a telephone circuit are:

(i) The transmitter, which converts sound waves into electrical impulses.
(ii) The receiver, which converts electrical impulses into sound waves.

A common form of transmitter is the carbon granule microphone. This type of transmitter has a diaphragm which vibrates whenever sound waves impinge upon it, the vibration of the diaphragm alternately compressing and releasing carbon granules which form the connection between the diaphragm and the rear contact. This has the effect of varying the resistance of the microphone in sympathy with the sound waves.

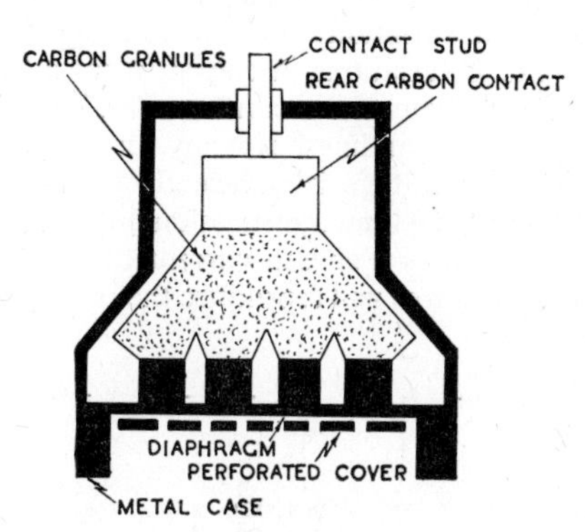

FIG. 9.15 Telephone Transmitter

(*b*) A common form of receiver has a mu-metal diaphragm which is held very close to the poles of a permanent magnet. The poles of the permanent magnet are fitted with coils so that whenever the current in the coils varies the resultant magnetic field strength also varies and so does the pull on the diaphragm. A current of one polarity will increase the pull on the diaphragm while a current of opposite polarity reduces the pull; hence if an alternating current is passed through the coil the diaphragm moves to and fro in step with the current.

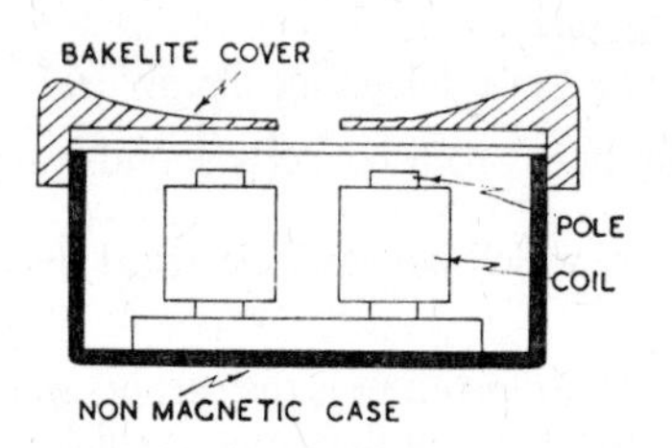

FIG. 9.16 Telephone Receiver

(*c*) Fig. 9.17 shows a circuit which can transmit sound in one direction only. The battery causes a current to flow through the carbon type transmitter and the low resistance primary of the induction coil. When a sound wave acts on the microphone the variations in resistance result in variations in the current flowing in the microphone circuit. These variations in current form a 'speech current' which induces a voltage in the secondary winding of the induction coil. As the secondary winding has many turns compared with the primary, the induced voltage is sufficiently high to overcome the effects of voltage drop if a long transmission line is used. The induced secondary voltage causes a speech current to flow in the receiver circuit and so the receiver will reproduce the original sound wave.

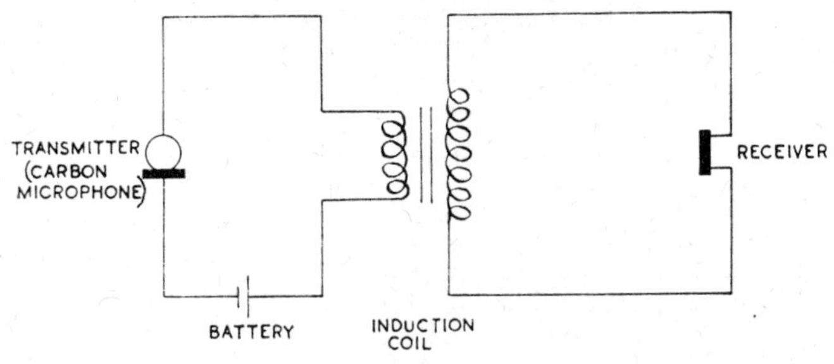

FIG. 9.17 Simple Telephone Circuit

(*d*) A more advanced type of circuit is required when it is necessary to provide two-way speech communication and a method of ringing. The circuit shown in Fig. 9.18 illustrates a simple method of meeting these requirements. In this circuit the calling switch, C, is held against the top contact by a spring, and the hook switch, H, is held down by the weight of the receiver. At this stage the line at each station is connected to a bell. If the call switch is pressed at station A, current flows from the battery, B2, to the bell at station B. Similarly, pressing the call-switch at B will ring the bell at A. Thus a caller at one station can attract the attention of a person at the other end. When the receivers at each station are lifted the hook switches rise and so make connection to the sound receiving and transmitting apparatus. Speech currents from either of the transmitters can now cause voltages to be induced in their respective induction coils which will operate the receivers at each end. A 'pressel' switch, P, is included in each microphone circuit to save current when the apparatus is not in use.

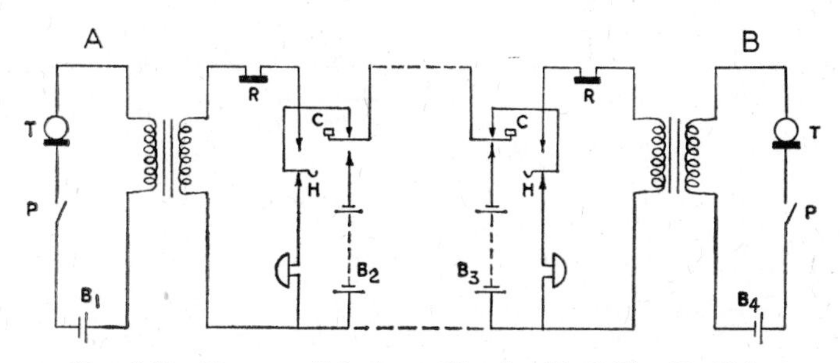

FIG. 9.18 Two-way Telephone Circuit with Calling Facilities

EXERCISES

The exercises which conclude this chapter are designed to illustrate the operation of simple types of alarm and communication systems.

EXERCISE No. 9.1 — To connect a continuous ringing bell and indicator board.

Materials

Continuous ringing bell.
4 Way indicator board.
4 Bell-pushes.
1/·036 Bell wire.
4·5V Dry battery.

Diagram

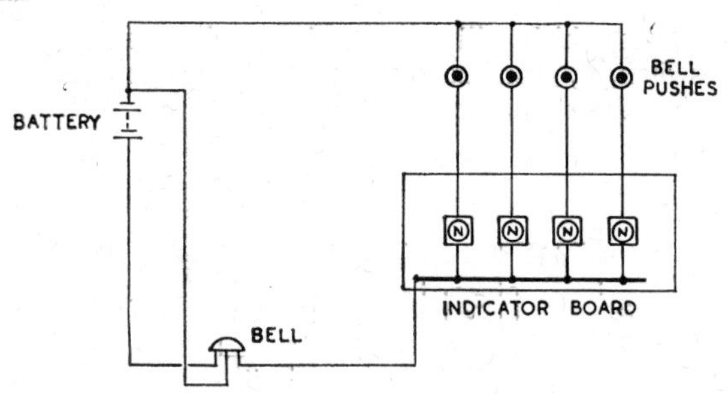

FIG. 9.19 Exercise 9.1

Procedure

1. Connect the circuit as shown in Fig. 9.19
2. Press each push in turn and observe that the bell rings and the appropriate indicator operates. Note that the bell continues to ring until reset.

Answer the following questions

1. With the aid of a sketch explain the operation of one of the indicator elements used in this circuit.
2. For what purposes could this circuit be used?

EXERCISE No. 9.2 — To connect a simple bell and relay circuit.

Materials

Trembler bell.
Bell relay.
Bell-push.
1/·036 Bell wire.
4·5V Dry battery.

Diagram

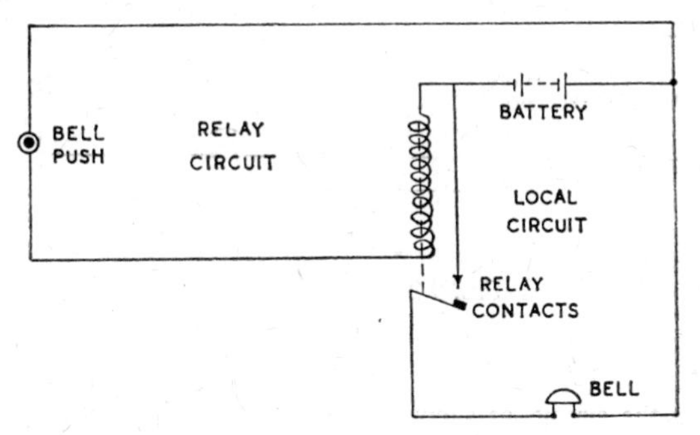

FIG. 9.20 Exercise 9.2

Procedure

1. Connect the circuit as shown in Fig. 9.20.
2. Press the push and observe the action of the relay.

Answer the following questions

1. Describe the operation of a relay.
2. In what circumstances would this circuit be used?

EXERCISE No. 9.3 **To connect a simple bell circuit using a bell transformer.**

Materials

Bell transformer.
Trembler bell.
Bell-push.
3/·029 Three-core T.R.S. or P.V.C. cable.
1/·036 Bell wire.

Diagram

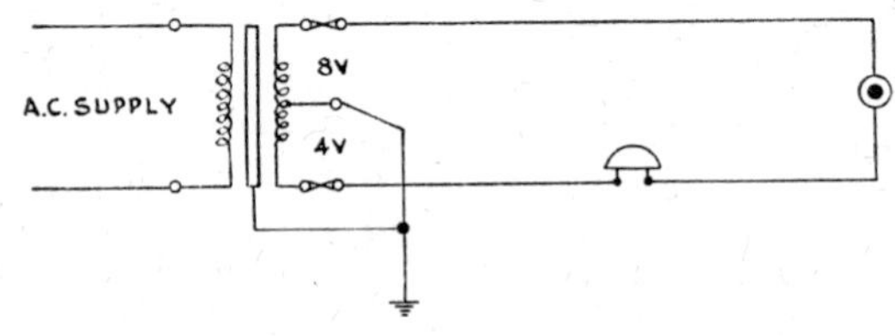

FIG. 9.21 Exercise 9.3

Procedure

1. Connect the circuit, as shown in Fig. 9.21, ensuring that the wiring from the a.c. supply to the primary of the transformer is carried out in 250V grade cable and that bell wire is used for the secondary wiring.
2. Switch on the a.c. supply and check that the circuit operates correctly.

Answer the following questions

1. Why should the transformer used to supply a bell circuit be of the 'double wound' type?
2. Why should the secondary winding of the transformer be earthed at one point?

EXERCISE NO. 9.4

To connect a closed circuit burglar alarm system.

Materials

Bell relay (with one pair of contacts 'normally open' and one pair normally closed').
Trembler bell.
Bell-push.
Single switch.
3 Normally closed alarm contacts.
1/·036 Bell wire.
4·5V Dry battery.

Diagram

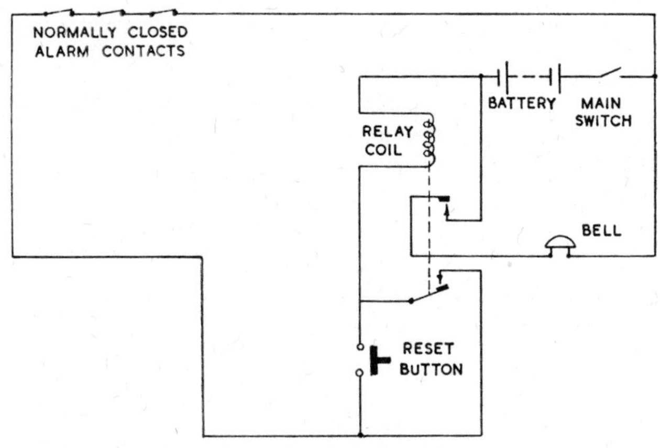

FIG. 9.22 Exercise 9.4

Procedure

1. Connect the circuit as shown in Fig. 9.22.
2. Switch on the main switch and note that the bell rings until the reset button is pressed.
3. Break one of the alarm contacts and note that the bell rings; close the contact again and note that the bell still continues to ring until the reset button is pressed.

Answer the following questions

1. What are the advantages of a closed circuit alarm system compared with an open circuit alarm system?
2. Carefully describe the operation of this circuit.

EXERCISE No. 9.5. To connect a simple telephone circuit.

Apparatus

2 Internal communication type telephones.
Dry batteries as required.

Procedure

1. Connect the telephones in accordance with the makers' instructions.
   N.B. – *It should be noted that the various makes of telephone differ slightly in their methods of connection, particularly in respect of the method of connecting the batteries.*
2. Check that the calling and speech circuits are functioning in a satisfactory manner.
3. Draw a circuit diagram for a simple telephone installation.

Answer the following questions

1. Explain with the aid of diagrams the action of a carbon type transmitter.
2. Explain with the aid of diagrams the action of a typical reciever.

CHAPTER 10

Secondary Cells

10.1

The two main types of secondary cell commonly used in electrical installations are:

(*a*) Lead-acid cell.
(*b*) Alkaline cell.

10.2 Lead-acid Cells

(*a*) The lead-acid cell consists of two sets of plates immersed in an electrolyte of dilute sulphuric acid. The plates are constructed in the form of lead grids which serve as a frame to support the 'active paste'. The paste, as applied to the plates by the manufacturer, is a mixture of lead oxides and sulphuric acid. This paste sets hard and during the initial (forming) charge is converted to lead peroxide on the positive plate and spongy lead on the negative plate.

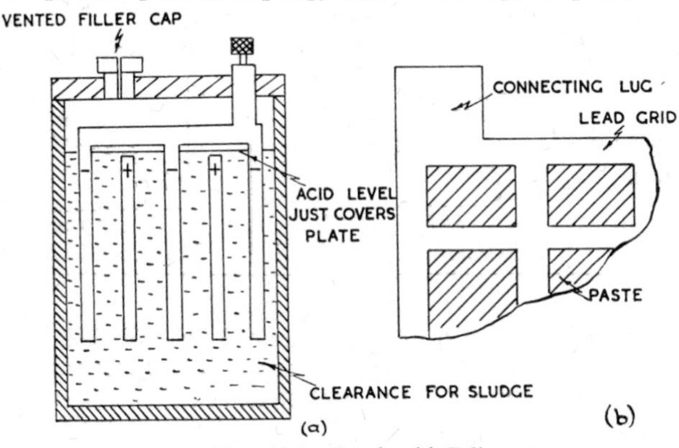

FIG. 10.1 Lead-acid Cell
(a) Arrangement of plates; (b) Detail of plates.

(*b*) The table below summarizes the changes which take place in a lead-acid cell during charging and discharging.

Charge and Discharge of Lead-acid Cell

Item	*Discharging*	*Charging*
Positive plates.	Tend to change to lead sulphate.	Change to lead peroxide.
Negative plates.	Tend to change to lead sulphate.	Change to spongy lead.
E.m.f.	Remains steady at approximately 2V per cell until the end of the useful discharge period after which it falls rapidly. It should not be allowed to fall below 1·8V per cell.	Rises to approximately 2V per cell.
Specific gravity.	Falls to approximately 1,100 at the end of the useful discharge period.	Rises to approximately 1,250.
Characteristic curves.	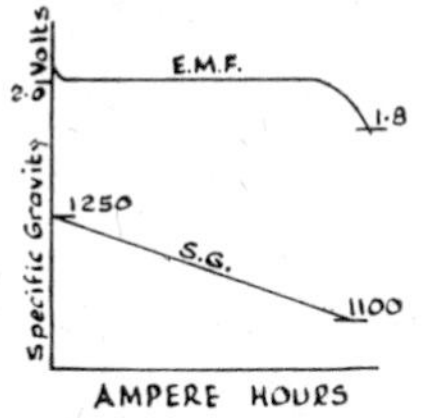 Volts 2·0 E.M.F. 1·8 Specific Gravity 1250 S.G. 1100 AMPERE HOURS DISCHARGE CURVE	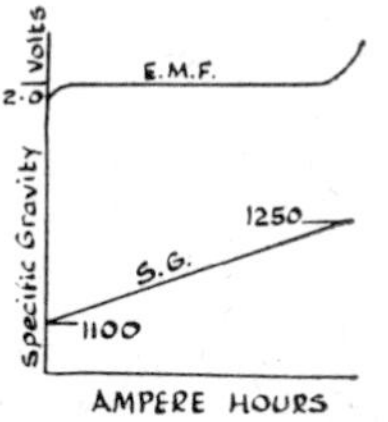 Volts 2·0 E.M.F. Specific Gravity 1250 S.G. 1100 AMPERE HOURS CHARGE CURVE

10.3 Installation Care and Maintenance of Lead-acid Cells

(*a*) Lead-acid cells tend to give off hydrogen when charged, due to electrolysis of the water in the electrolyte. A mixture of hydrogen and air in certain proportions is highly explosive so the following precautions must be observed in all rooms where cells are installed or charged:

(i) The room must be well ventilated so that the hydrogen may escape and not form a dangerous concentration.
(ii) No naked lights, or smoking, must be allowed in the battery room.

(*b*) The fumes from lead-acid cells are very corrosive so any metal work in the vicinity of the cells should be protected by painting with an acid resisting bitumastic paint. The terminals of cells should be kept lightly coated with petroleum jelly (or a suitable proprietory compound) in order to prevent corrosion, and the outside of the cell should be kept clean and dry.

(*c*) Lead-acid cells require regular maintenance if they are to be kept in an efficient working condition. The electrolyte must be maintained at the correct level by the addition of distilled water when required. Ordinary tap water should never be used for topping up, neither should acid, except in the special circumstances stated in paragraph 10.3(d) below. Cells should be charged regularly; if a cell is allowed to remain discharged for a long period the plates become covered with a hard white deposit of lead sulphate which causes permanent damage to the cell; this is known as 'sulphation'. The state of charge of a cell can be checked either by using a hydrometer, which measures the specific gravity of the electrolyte, or by using a voltmeter. The voltage of a cell should never be allowed to drop below 1·8V per cell as this indicates that the useful charge of the cell is exhausted.

(*d*) When sulphuric acid is mixed with water a considerable amount of heat is evolved. If distilled water is poured into concentrated acid, the heat produced is sufficient to cause dangerous spitting of the acid, but if the acid is poured a little at a time into the water, the heat is evolved more slowly and safely. Thus when preparing electrolyte care must be taken ALWAYS to add acid to water and NEVER water to acid. The electrolyte must be allowed to cool before its specific gravity is checked. It is not normally necessary to use acid when topping up cells; the principal loss of electrolyte is due to evaporation of the water and so topping up with distilled water is usually all that is required. Nevertheless if there is reason to suspect that the acid in a cell is unduly weak, this can be checked by first charging the cell until it has been gassing freely for some time so ensuring that the cell is fully charged, and then testing the acid strength with a hydrometer. If a low reading is obtained, acid with a strength somewhat greater than that normally used in the cell may be added a little at a time, until the desired specific gravity reading is obtained. Care should be taken to ensure thorough mixing of the acid with the existing electrolyte so that false readings of specific gravity are not obtained. Because of the heat that is evolved when adding acid to the electrolyte acid should only be added a little at a time to the cell, sufficient time being allowed for the cell to cool before more acid is added.

10.4 Alkaline Cells

(*a*) A common type of alkaline cell is the nickel cadmium cell. This cell consists of two sets of plates immersed in an electrolyte of a solution of potassium hydrate (caustic potash) in distilled water. The plates are constructed in the form of perforated steel compartments which contain the active materials; the positive plates contain a mixture of nickel hydroxide and graphite while the negative plates contain a mixture of cadmium and iron oxides.

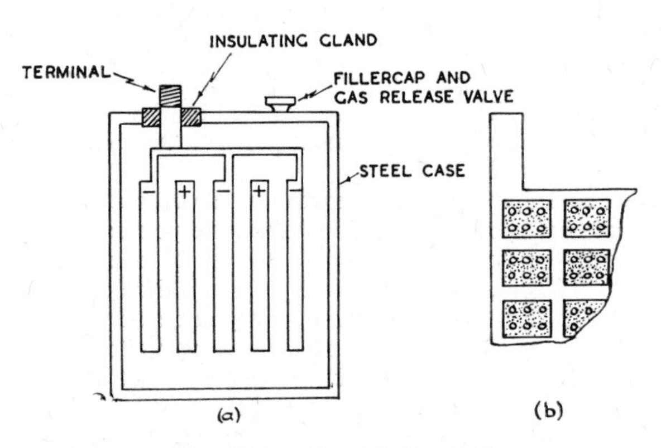

FIG. 10.2 (a–b) Alkaline Cell
(a) Arrangement of plates; (b) Detail of plates.

(*b*) The tables below summarize the changes which take place in an alkaline cell while charging and discharging, and compare the properties of lead acid and alkaline cells.

Charge and Discharge of Alkaline Cell

Item	*Discharging*	*Charging*
Positive plates.	Lose oxygen.	Restored to original condition.
Negative plates.	Oxidize.	Restored to original condition.
E.m.f.	Usually falls to 1V at end of useful discharge.	Rises to approximately 1·4V
Specific gravity.	Remains constant at approximately 1,170.	Remains constant at approximately 1,170.
Characteristic curves.	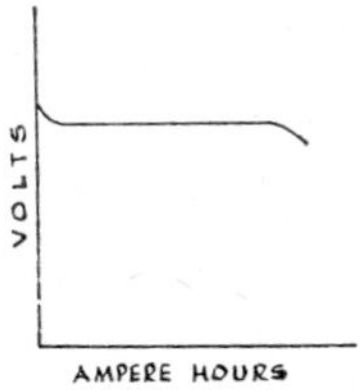 VOLTS / AMPERE HOURS DISCHARGE CURVE	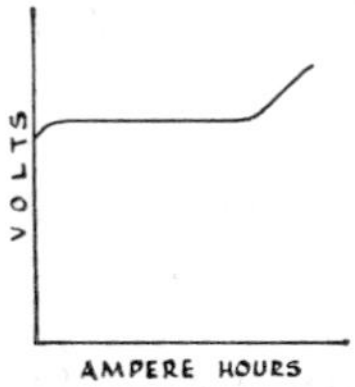 VOLTS / AMPERE HOURS CHARGE CURVE

Comparison of Lead-acid and Alkaline Cells

Item	*Lead-Acid Cell*	*Alkaline Cell*
Average e.m.f.	2·0V	1·2V
Efficiency (ampere hour).	85 to 90%.	75 to 80%.
Efficiency (watt hour).	70 to 75%.	60 to 65%.
Cost.	An expensive source of electrical energy.	Even more expensive than the lead-acid cell.
Strength.	Adversely affected by vibration, and heavy currents.	Withstands vibration and is capable of heavy discharge currents.

10.5 Battery Charging

(*a*) Batteries are charged by passing direct current through them from positive to negative terminals. The cells are normally connected in series so that each cell receives the same charging current. If the cells were connected in parallel it would be difficult to maintain the correct division between currents, since cells having a slightly higher e.m.f. would tend to discharge into those cells possessing a lower e.m.f.

(*b*) When using the constant voltage method the charging voltage is maintained constant at a value slightly in excess of the e.m.f. of a fully charged cell multiplied by the number of cells connected in series. The charging current using this method depends on the difference between the charging voltage and the total e.m.f. of the cells; this current may be calculated from the formula:

$$I = \frac{V - E}{R}$$

Where V is the charging voltage
E is the total e.m.f. of the cells
R is the total internal resistance.

The above formula indicates that at the commencement of the charge, when the e.m.f. of the cells is low, the charging current will be large; the charging current will gradually reduce in value as charging progresses. Fig. 10.3 shows the circuit of a constant voltage battery-charger; the ballast resistor, R, which has a low value helps to prevent excessive currents flowing at the commencement of the charge.

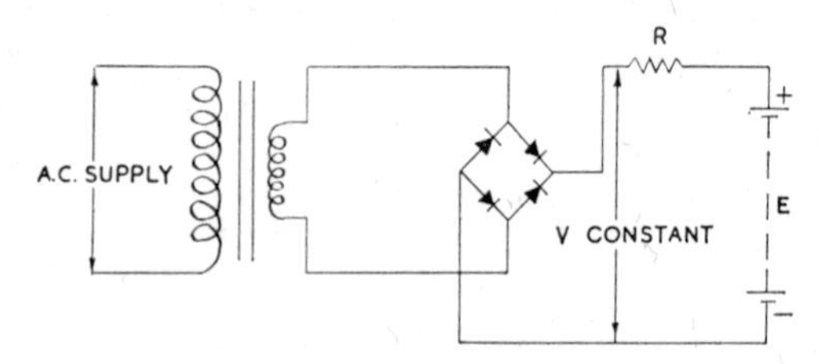

FIG. 10.3 Constant Voltage Battery Charger

(*c*) When using the constant current method, the charging current is kept constant at a suitable value for the cells concerned. Fig. 10.4 (a) shows how this may be done using a rheostat in series with the output of the charger. The current flowing through the rheostat depends on its setting and on the p.d. between its terminals, this p.d. is the difference between the output voltage of the rectifier and the voltage required to charge the cells. Thus if the rectifier output voltage is much higher than the actual charging voltage needed for the cells, the percentage change in the p.d. across the rheostat as the cells charge up is small and the current remains reasonably constant. An alternative method is to control the current by using a variable inductor connected in series with the a.c. input to the charger as shown in Fig. 10.4(b).

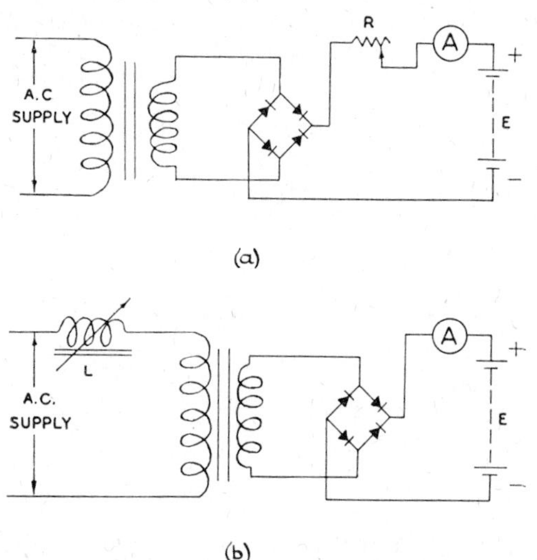

FIG. 10.4 (a–b) Constant Current Battery Chargers
(a) Rheostat control; (b) Inductor control.

EXAMPLE No. 1

Thirty lead-acid secondary cells are to be charged at constant voltage. The e.m.f. of each cell at the beginning and end of charge is 1·9V and 2·7V respectively, and the internal resistance of each cell is 0·1 ohms. Calculate:

(*a*) The minimum charging voltage required.
(*b*) The initial charging current.

SOLUTION

(*a*) Minimum charging voltage = number of cells × final voltage of cell

$$V = 30 \times 2{\cdot}7$$
$$V = 81V$$

(*b*)
$$I = \frac{V - E}{R}$$
$$I = \frac{81 - (30 \times 1{\cdot}9)}{30 \times 0{\cdot}1}$$
$$I = 8A$$

EXAMPLE No. 2

Thirty lead-acid secondary cells are to be charged using the constant current method. The e.m.f. of each cell at the beginning and end of charge is 1·9V and 2·7V respectively. The output voltage of the rectifier used in the charger is 157V.

Ignoring the internal resistance of the cells, calculate:

(*a*) The value in ohms to which the rheostat must be set to give an initial charging current of 5A.

(*b*) The current at the end of the charge if the rheostat is left at this setting.

SOLUTION

(*a*) P.D. across rheostat at commencement of charge

$$= 157 - (30 \times 1{\cdot}9)$$
$$= 100V$$

Rehostat setting, in ohms: $R = \frac{V}{I}$

$$R = \frac{100}{5}$$
$$\underline{R = 20\Omega}$$

(*b*) P.D. across rheostat at end of charge $= 157 - (30 \times 2{\cdot}7)$

$$= 76V$$

Current at end of charge: $I = \frac{V}{R}$

$$I = \frac{76}{20}$$
$$\underline{I = 3{\cdot}8A}$$

EXERCISES

The two exercises which conclude this chapter are designed to give practice in maintaining lead-acid cells and to illustrate the action of a battery-charger.

Exercise No. 10.1 — Maintenance of lead-acid cells.

Apparatus

12V Battery-charger.
6 2V Lead-acid cells (partially discharged).
1 Cell-testing voltmeter.
1 Hydrometer.
Distilled water.
Petroleum jelly.

Procedure

1. Check the voltage and specific gravity of each cell.
2. Check the level of the electrolyte in the cells and add distilled water if necessary.
3. Connect the 6 2V lead-acid cells in series, taking care to observe correct polarity.
4. Connect the positive terminal of the battery-charger to the positive terminal of the first cell and the negative terminal of the battery-charger to the negative terminal of the last cell; remove all vent plugs and switch on the battery-charger.
5. After some time (when the cells are gassing freely) switch off the battery-charger. Check the specific gravity and terminal voltage of each cell.
6. Disconnect the cells, replace the vent plugs, and wipe off any moisture from the cell tops; smear a light film of petroleum jelly over the terminals.

Answer the following questions

1. What values of specific gravity and terminal voltage are to be expected for full-charged cells?
2. Why should no naked flames be allowed in the vicinity of batteries that are being charged?
3. Why should lead-acid cells be recharged at regular intervals?

EXERCISE No. 10.2 — To investigate a single-phase a.c. battery-charger.

Apparatus

Small battery-charger suitable for charging 12V battery.
1 m.c. voltmeter 0–20V.

Procedure

1. Remove the covers from battery-charger, examine the components, and trace the circuit connections.
2. Connect the charger to the a.c. supply and switch on.
3. Connect the voltmeter to the secondary side of the transformer and note that the instrument will not read.
4. Connect the voltmeter to the output terminals and note the reading.
5. Draw a neat diagram of the battery-charger circuit.

Answer the following questions

1. Is the battery-charger investigated of the constant current or constant voltage type?
2. Explain with the aid of diagrams, the meaning of rectification of an alternating current.
3. Why was no reading obtained on the voltmeter when connected to the transformer secondary?

CHAPTER 11

D. C. Machines

11.1

There are two types of d.c. machine, the dynamo, or d.c. generator, which converts mechanical energy into electrical energy; and the motor, which converts electrical energy into mechanical energy. Both generators and motors rely for their operation on two fundamental electro-magnetic effects, the dynamo and the motor effect.

11.2 Dynamo Effect

Whenever a conductor moves through a magnetic field an e.m.f. is induced in the conductor. The magnitude of the induced e.m.f. depends on the strength of the magnetic field and the speed at which the conductor moves. The direction of the induced e.m.f. can be found by using the 'right hand' rule as follows: Hold the right hand flat so as to receive the magnetic flux on the palm with the thumb extended and pointing in the direction that the conductor moves, then the fingers point in the direction of the induced e.m.f.

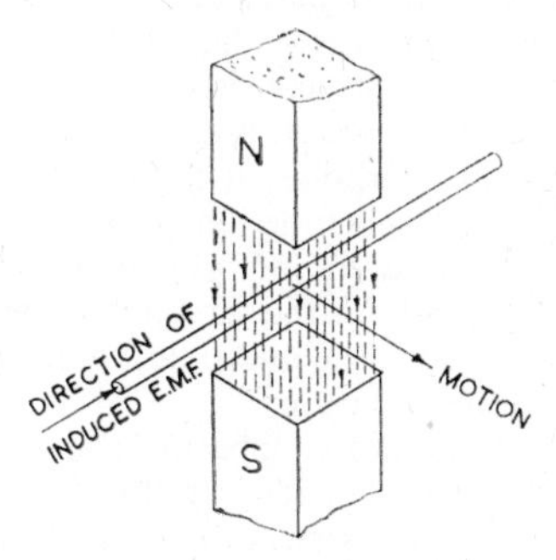

FIG. 11.1 Induced e.m.f.

11.3 Motor Effect

Whenever a conductor, situated in a magnetic field, carries a current, it experiences a force. The magnitude of this force depends on the strength of the magnetic field and the value of the current in the conductor. The direction of the force can be found by using the 'left hand' rule as follows: Hold the left hand flat so as to receive the magnetic flux on the palm with the fingers pointing in the direction of the current and the thumb extended. Then the thumb points in the direction of the force.

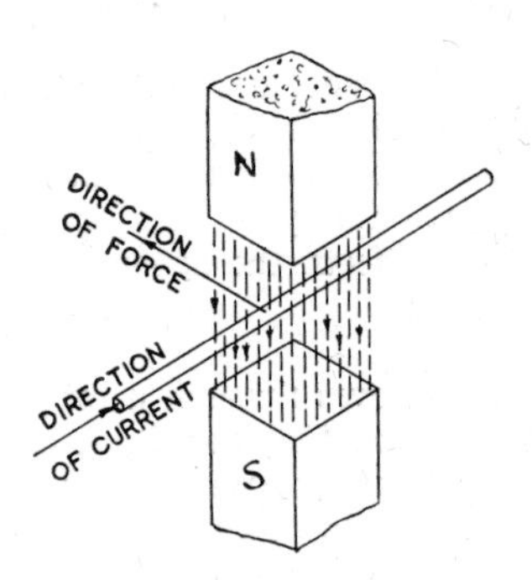

FIG. 11.2 Force on a Conductor

11.4 Construction of d.c. Machines

A d.c. machine has three main parts, the field system, the armature and the commutator with its associated brush gear.

(*a*) The field system usually consists of an electro-magnet with one or more pairs of poles which are normally fixed to the outer frame of the machine. Interpoles, if fitted, are small poles fitted between the main poles; their function is to reduce sparking at the commutator. The polarity of an interpole in a dynamo must be the same as the next main pole in the direction of rotation, and vice versa in the case of motors The interpole windings are always connected in series with the armature.

(*b*) The armature is the revolving part of the machine and usually consists of a laminated iron cylinder with slots equally spaced around its circumference. Coils are fitted into the slots to form the armature winding.

(*c*) The commutator is made from copper bars insulated from each other by mica. The armature coils are connected to the commutator bars and so make contact with the carbon brushes bearing on the commutator. The result is that the alternating e.m.f. induced in the armature coils is converted into a direct e.m.f. at the brushes.

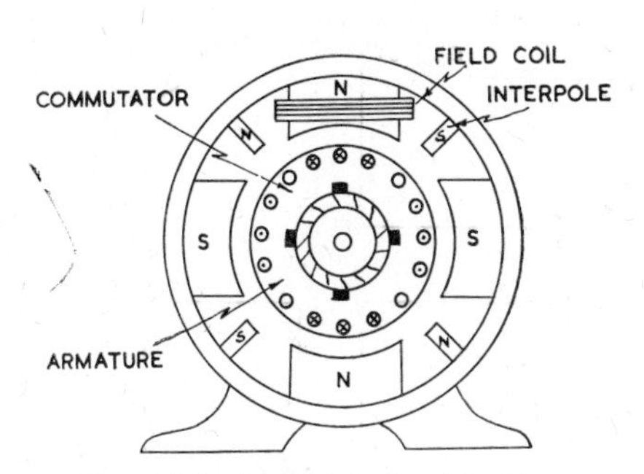

FIG. 11.3 Four Pole d.c. Machine

(*d*) Types of d.c. machine are named according to the way in which the field windings are connected. A machine in which the field coils are connected in series with the armature is called a series machine, and a machine in which the field coils are connected in parallel with the armature is called a shunt machine. A compound machine has two sets of field coils one set connected in series and the other in parallel with the armature. A separately excited machine has its field coils energized by a separate d.c. supply. Fig. 11.4 shows the electrical connections for the various types of d.c. machine (both dynamos and motors). The method of field connection used in a d.c. machine is the main factor which determines its operating characteristics, the characteristics of the types of dynamos and motors in common use are discussed in paras. 11.5 and 11.6 below.

11.5 Characteristics of Dynamos

(*a*) The behaviour of any type of dynamo can be deduced from the following relationships:

(i) The generated e.m.f. is directly proportional to the speed.

(ii) The generated e.m.f. is directly proportional to the magnetic flux per pole.

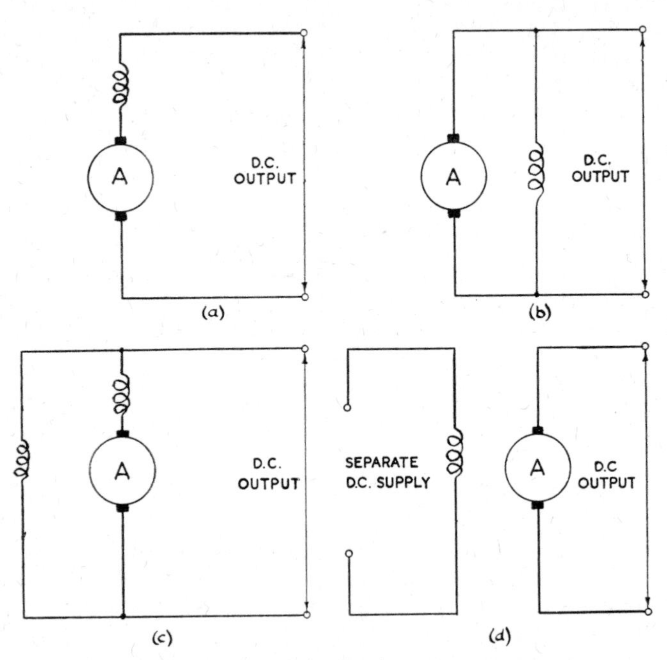

FIG. 11.4 (a–d) Connections of d.c. Machines
(a) Series; (b) Shunt; (c) Compound; (d) Separately excited.

(iii) The magnetic flux per pole is approximately proportional to the current in the field winding, for practical values of field current.

(iv) The terminal voltage is given by:

$$V = E - I_a (R_a + R_s) - V_b$$

where E = generated e.m.f.
I_a = armature current.
R_a = armature resistance.
R_s = resistance of series field
V_b = brush contact drop.

(*b*) In a series dynamo the only current flowing in the field windings is the load current so when the current supplied by the dynamo is small the current flowing in the field coils is also small, which results in a weak magnetic field, and a small generated e.m.f. At a higher value of load current the magnetic field is strengthened giving a higher e.m.f. Thus the e.m.f. increases as the current output increases in the manner illustrated by Fig. 11.5. The characteristics of this type of dynamo limit its use to such applications as a 'booster' to offset the voltage drop in a long cable.

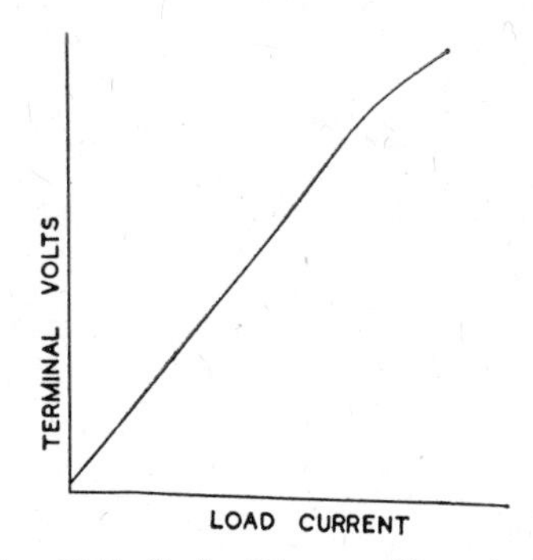

FIG. 11.5 Series Dynamo Characteristic

(*c*) In a shunt dynamo the field winding is connected in parallel with the armature and so the field current is determined by the terminal voltage, and not by the load current taken from the dynamo. The terminal voltage of this type of dynamo tends to be fairly constant but drops slightly as the load current increases. This voltage drop occurs in the first instance because of the resistance of the armature windings but, as any fall in terminal voltage must reduce the field current, which in turn reduces the generated voltage, the overall voltage drop is greater than that caused by the armature resistance alone. The characteristics of this type of dynamo make it suitable for the majority of applications where a fairly steady voltage is required such as small lighting plants, battery-chargers, etc. The terminal voltage can be easily controlled by connecting a rheostat (field regulator) in series with the field winding. When the field current is reduced (by increasing the rheostat's resistance) the generated voltage is also reduced.

(*d*) A compound dynamo has both shunt and series field windings. The shunt field winding provides most of the magnetization of the

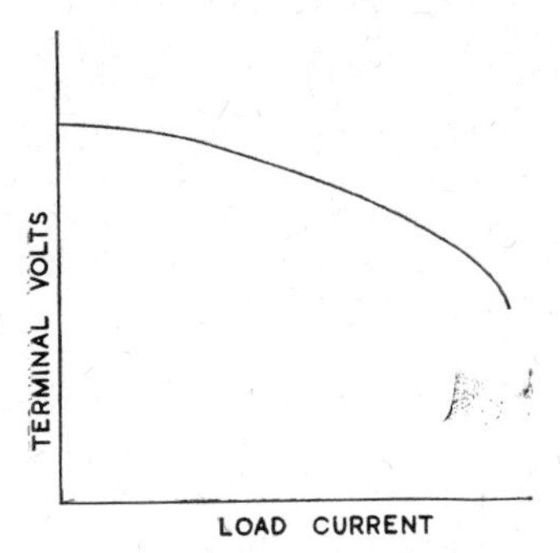

FIG. 11.6 Shunt Dynamo Characteristic

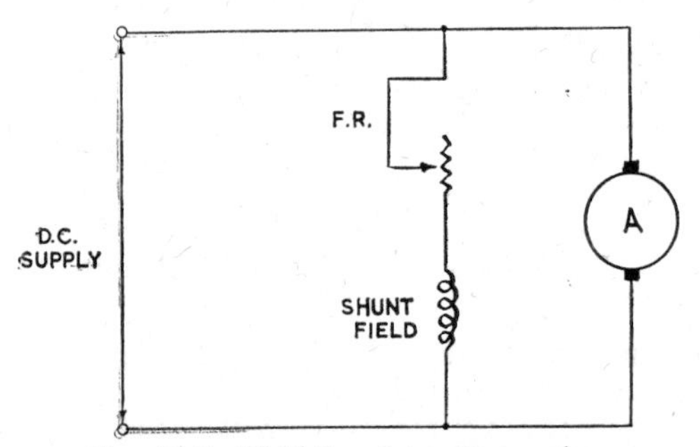

FIG. 11.7 Field Regulator Connections

poles but, as the load current increases, the effect of the series winding is to increase the pole strengths. This has the effect of increasing the generated voltage as the load rises so tending to compensate for the voltage drop in the armature resistance. A machine with an almost constant terminal voltage is known as a 'level' compounded type. A machine giving an increase of terminal voltage with load current is known as an 'over-compounded' type. The characteristics of the level compounded type of dynamo make it suitable for applications where a more constant voltage than that provided by a shunt dynamo is needed. Over-compounded types are useful where a load is to be supplied via a long cable, as the rise in voltage at the dynamo terminals helps to offset the voltage drop in the cable as the load current increases.

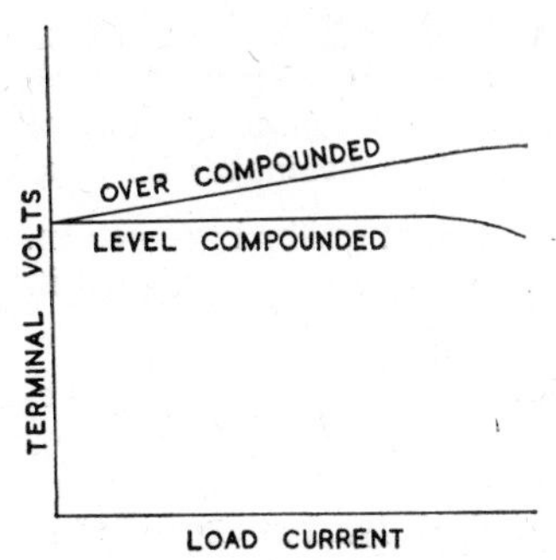

FIG. 11.8 Compound Dynamo Characteristic

(*e*) The output voltage of the separately excited dynamo is determined mainly by the voltage of the separate d.c. supply to the field, and so can be varied smoothly over a wide range; the polarity of the generated voltage can be reversed by reversing the field current. The terminal voltage tends to fall slightly on load due to the effect of armature resistance.

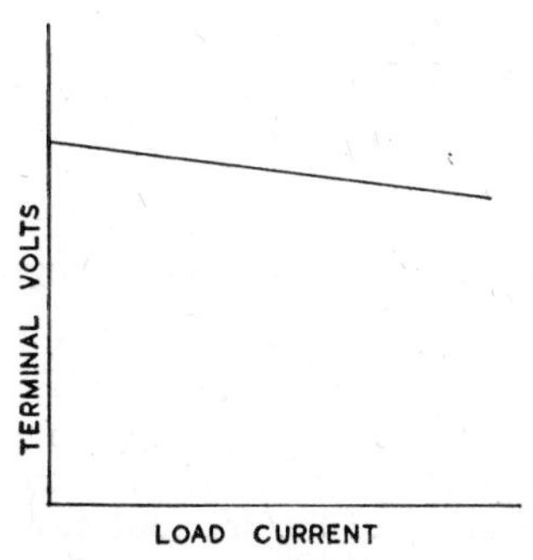

FIG. 11.9 Separately Excited Dynamo Characteristic

(*f*) All self-excited dynamos whether series, shunt or compound rely upon residual magnetism to provide their initial excitation. For example, in the case of a shunt dynamo, when the machine is first started, the residual magnetism of the poles causes a small e.m.f.

to be induced in the armature. This in turn causes current to flow in the field winding which strengthens the poles so increasing the e.m.f. The effect is progressive until the machine attains full voltage. It should be noted that the machine can only excite if:

(i) There is some initial residual magnetism.
(ii) The field winding is connected so that it tends to strengthen the initial magnetism in the poles.

If a dynamo has lost its residual magnetism and so fails to excite, the leads from a battery may be momentarily touched on the dynamo terminals to provide enough magnetism to initiate excitation. The polarity of a self-excited dynamo depends on the polarity of the residual magnetism, it cannot be reversed merely by reversing the field connections as this would simply result in the dynamo failing to excite.

11.6 Characteristics of Motors

(*a*) When the armature of a motor is rotating it acts as a dynamo and generates an e.m.f. This e.m.f. opposes the flow of current through the armature and is therefore called a 'back e.m.f.'. The back e.m.f. is important because it is the main factor which controls the flow of current through the armature, the armature current being given by the formula:

$$I_a = \frac{V - E_b}{R_a}$$

Where I_a = armature current.
V = supply voltage.
E_b = back e.m.f.
R_a = armature resistance.

The value of the back e.m.f. is proportional to the speed and to the magnetic flux per pole, as is the e.m.f. generated by a dynamo, and since in any practical motor the armature resistance is small, the back e.m.f. is always only slightly less than the supply voltage.

(*b*) The torque produced by a motor is proportional to the field current and to the magnetic flux per pole. If the torque produced by a motor is more than that required to drive the load it will accelerate, so increasing the back e.m.f. and reducing the armature current, until eventually just enough current is taken to produce the load torque. If on the other hand a motor produces less torque than that required by the load it will slow down, thus reducing the back e.m.f. and allowing more current to flow through the armature until once again the torque produced by the armature current balances the load torque. The current consumed by any motor will be that required to produce the load torque, while the speed will automatically adjust itself to the value required to give the correct back e.m.f. If the magnetic field of a motor is weakened the motor must run faster to maintain its back e.m.f., and if the magnetic field is strengthened the motor will slow down.

(*c*) In a series motor the armature current also passes through the field winding so the magnetic field is strong when the motor is heavily loaded, giving rise to a high torque at a low speed. If the load is reduced the magnetic field is weakened so the speed increases; at very light loads the speed can increase to a dangerously high value. The characteristics of this motor make it suitable for applications where a high starting torque is required, and where the speed variation with load is no disadvantage. The motor must always be directly coupled to the load as the speed would become excessive if the load were suddenly removed. This makes the motor ideal for traction purposes, a further advantage in this case being that the high starting torque is still available even if there is considerable voltage drop in the supply mains. Small series motors will operate from a.c. supplies as well as d.c.; these are sometimes known as universal motors and are used to drive vacuum cleaners, and similar domestic electrical appliances.

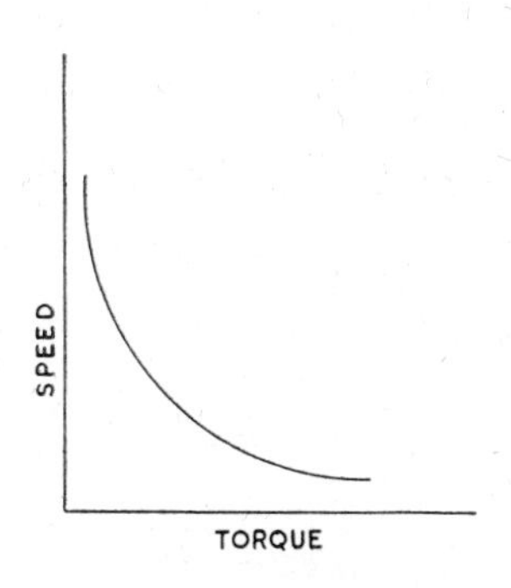

FIG. 11.10 Series Motor Characteristic

(*d*) In a shunt motor the field winding receives the full supply voltage and so the strength of the magnetic field is constant, hence the speed is fairly constant, falling slightly at high loads due to the effect of armature resistance. The speed can be easily controlled using a field rheostat as explained in para. 11.7 below. The characteristics of this motor make it suitable for situations where a fairly constant speed is required such as driving machine-tools, line shafting, etc. It is also useful where a moderate degree of speed control is required.

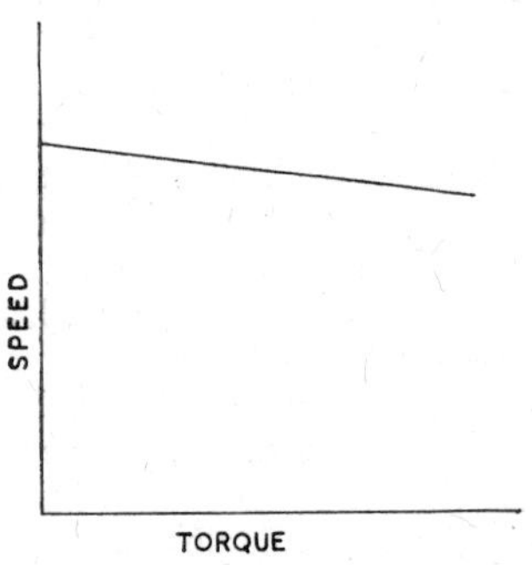

FIG. 11.11 Shunt Motor Characteristic

(*e*) The compound motor has both shunt and series field windings and so its characteristics are intermediate between the series and shunt types, giving a high starting torque together with a safe no-load speed. These factors make it suitable for use with heavy intermittent loads such as lifts, hoists, fly presses, etc.

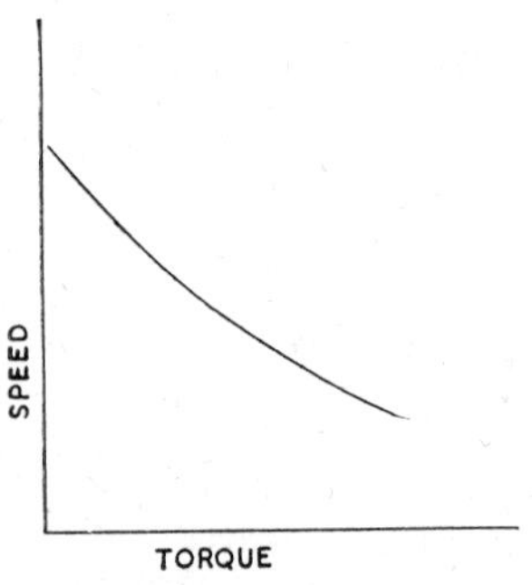

FIG. 11.12 Compound Motor Characteristic

11.7 Speed Control

(*a*) In general there are two methods of controlling the speed of d.c. motors, these are:

(i) By varying the voltage applied to the armature.
(ii) By varying the strength of the magnetic field.

A rheostat connected in series with the armature will reduce the voltage applied to the armature, thereby lowering the speed. Unfortunately this method of speed control has the following serious disadvantages:

(i) The rheostat must be capable of carrying the full armature current, and so tends to be expensive.
(ii) The power loss in the rheostat lowers the overall efficiency.
(iii) The effectiveness of the rheostat varies with the load current.

(*b*) A widely used method of controlling the speed of shunt and compound motors is to connect a rheostat in series with the shunt field as shown in Fig. 11.13. When the rheostat is adjusted so as to increase its resistance the field current is decreased, so weakening the magnetic field. The motor now has to run faster to maintain the back e.m.f. It is necessary to select the motor so that its speed, with the rheostat at its minimum setting, is the lowest that may be required, as an increase in rheostat resistance only increases the speed. A speed variation of approximately 3 to 1 can be obtained with average machines without weakening the poles unduly; a wider speed range usually calls for a specially constructed machine.

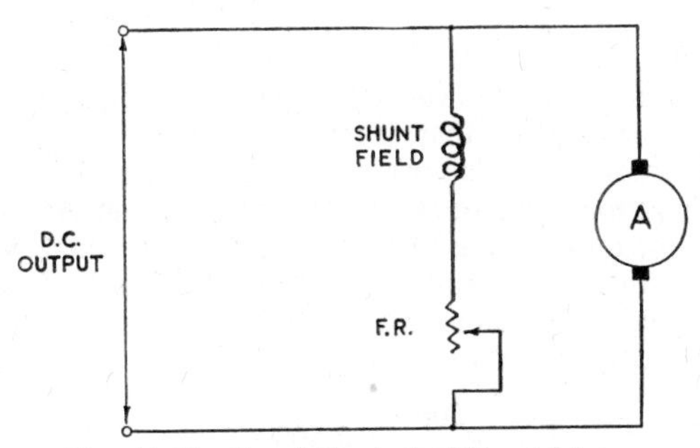

FIG. 11.13 Speed Control of Shunt Motor

(*c*) The speed of a series motor may be regulated by using divertor resistors connected as shown in Fig. 11.14. The field divertor reduces the current in the series field so increasing the speed of the motor, while the armature divertor increases the current in the series field thereby reducing the speed of the motor.

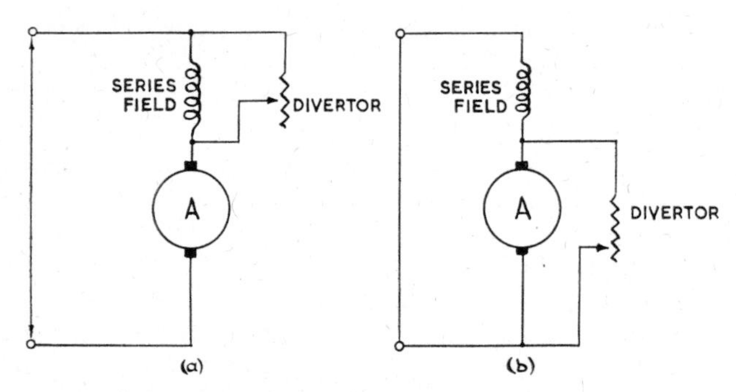

FIG. 11.14 Speed Control of Series Motor
(a) Field divertor; (b) Armature divertor.

11.8 Reversal of d.c. Motors

In order to reverse the direction of rotation of any d.c. motor it is necessary to reverse the connections of either the field winding or the armature but not both. Fig. 11.15 shows the alterations in connections required for reversed rotation of series, shunt and compound motors.

11.9 Motor Starting

(*a*) When a motor is first switched on there is no back e.m.f. to limit the armature current so that, if no method of limiting the starting current is employed, this current often becomes large enough to cause damage to the motor and its associated wiring. Also, excessive voltage drops may be caused in the supply mains thereby affecting other apparatus. For example, if a 500V d.c. motor having an armature resistance of 0·02 ohms is connected directly to the supply then the initial starting current will be:

$$I = \frac{500}{0{\cdot}02} = 2{,}500\text{A}$$

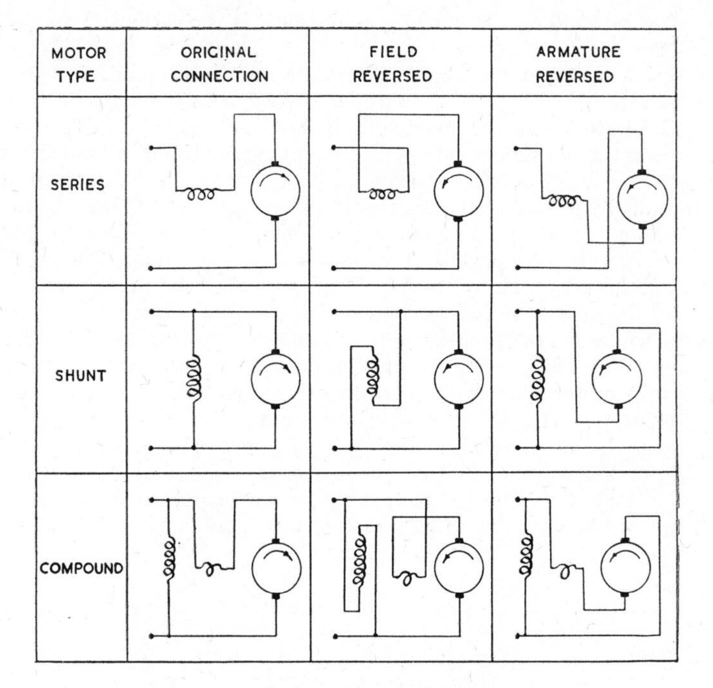

FIG. 11.15 (a–i) Reversal of d.c. Motors

A good method of reducing the starting current is to connect a rheostat in series with the armature, the rheostat being gradually cut out of circuit as the motor gathers speed so developing a back e.m.f.

(*b*) As well as being able to limit the starting current, the motor starter should include safety devices to:

(i) Automatically re-connect the starting resistors whenever the motor is stopped.
(ii) Prevent automatic re-starting should the motor stop owing to supply failure.
(iii) Automatically disconnect the motor in the event of a sustained overload.

(*c*) A good example of a d.c. motor starter incorporating simple devices fulfilling the above requirements, is the face-plate starter for a shunt motor, shown in Fig. 11.16. The starter arm (S.A.) is spring loaded to return to the OFF position and when in this position the supply to the motor is interrupted. When the starter arm is moved sufficiently to make contact with the first stud the field winding receives the full supply voltage but the armature current is limited by the starting resistors. As the motor gathers speed the starter arm is gradually moved to the ON position so cutting out the starting resistors from the armature circuit. The starter arm is held in the ON position by the magnetic attraction of the no-volt release (N.V.R.) which is energized by the field current. Should the supply voltage fail the no-volt release becomes de-energized and the starter arm returns to the OFF position under the influence of its spring. Thus the no-volt release provides a means of preventing automatic restarting of the motor when it has inadvertently stopped due to a failure of the supply. The no-volt release also provides protection against an open circuit occurring in the field winding. All the motor current flows through the coil of the overload relay (O.L.R.) and if this current becomes excessive, for example due to overloading the motor, the relay armature is attracted, so short-circuiting the

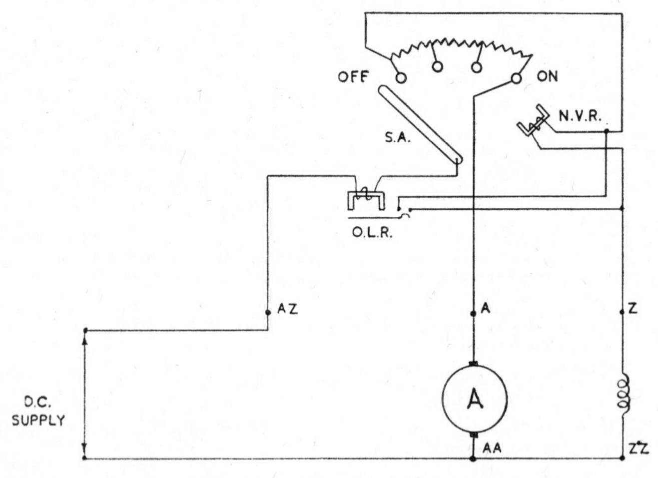

FIG. 11.16 Face-Plate Starter

[illegible]o-volt release. This again allows the starter arm to return to the [illegible] position. When stopping the motor it is good practice to switch [illegible] by means of the isolator or main switch rather than to pull back the starter arm, as this can result in burning the starter contacts.

EXERCISES

The exercises which conclude this chapter are designed to illustrate the operation of d.c. machines.

EXERCISE NO. 11.1 **To investigate the starting current of a d.c. motor.**

Apparatus

Shunt motor.
Face-plate starter.
Ammeter. (The ammeter range should be at least three times the full load current of the motor.)

Diagram

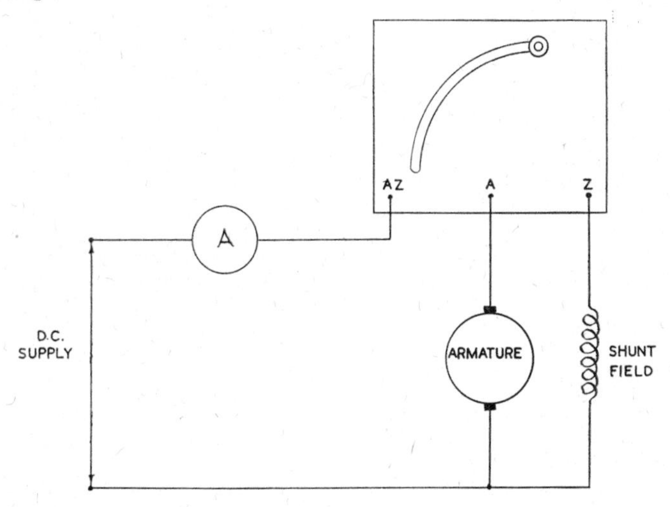

FIG. 11.17 Exercise 11.1

Procedure

1. Connect the circuit as shown in Fig. 11.17.
2. Switch on supply, move the starter arm to the first stud and observe that the ammeter shows a high reading which immediately falls as the motor starts to rotate.
3. As the motor gathers speed move the starter arm steadily to the ON position, noting that there is a sudden rise in the current as each section of starting resistance is cut out, and that the current falls as the speed rises.

Answer the following questions

1. Define back e.m.f.
2. How is the starting current of a d.c. motor affected by the back e.m.f.?
3. What are the requirements of the I.E.E. Regulations concerning:

   (*a*) The sizes of cables in motor circuits, and
   (*b*) The rating of fuses protecting motor sub-circuits?

EXERCISE No. 11.2 — To connect a d.c. shunt motor, starter, and field regulator.

Apparatus

Shunt motor.
Face-plate starter.
Field regulator.
Main switch.

Diagram

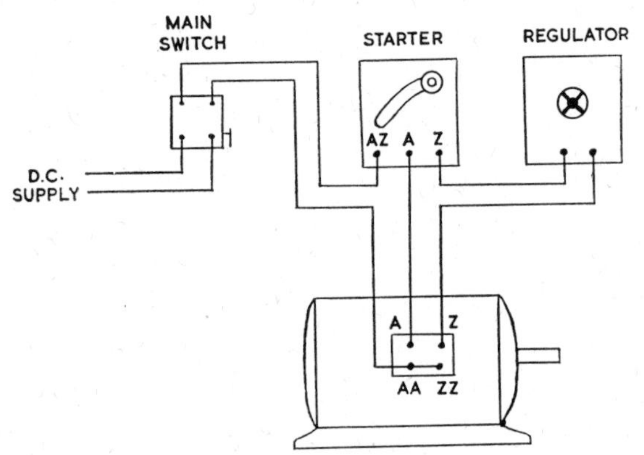

FIG. 11.18 Exercise 11.2

Procedure

1. Connect the circuit as shown in Fig. 11.18; start the motor and note the direction of rotation.
2. Switch off; reverse the armature connections at the motor terminal block, start the motor and again note the direction of rotation.
   Make a neat diagram of a 'face-plate' motor starter.

Answer the following questions

1. What are the requirements of the I.E.E. Regulations concerning:
   (*a*) Efficient means of isolation.
   (*b*) Prevention of automatic restarting.
   (*c*) Means of protection against excess current for a motor exceeding $\frac{1}{2}$ horse-power?
2. What means are provided in the face-plate starter to comply with the above regulations?

EXERCISE No. 11.3

Speed control of a d.c. shunt motor.

Apparatus

Shunt motor.
Face-plate starter.
Field regulator.
Tachometer.

Diagram

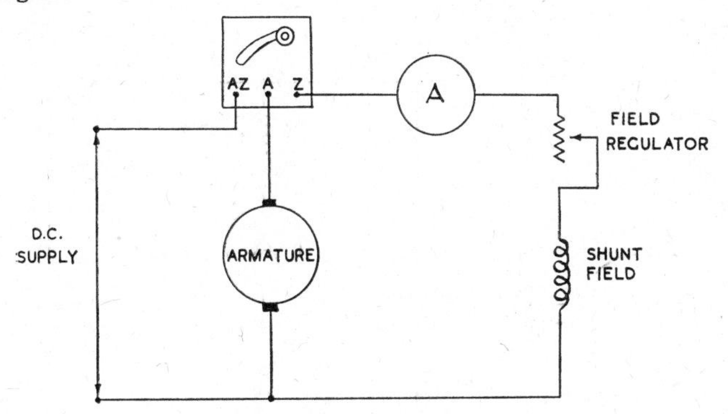

FIG. 11.19 Exercise 11.3

Procedure

1. Connect the circuit as shown in Fig. 11.19.
2. Start the motor, adjust the field regulator to give maximum field current and note the current and motor speed.
3. Reduce the field current by a small amount and again note current and speed.
4. Repeat step 3, reducing the field current in small steps until the minimum field current is achieved.
   Plot a graph of speed on a base of field current.

Answer the following questions

1. Explain carefully the effect of connecting a rheostat in series with the field winding of a d.c. shunt motor.
2. Why could an open circuit in the field winding of a d.c. shunt motor give rise to dangerous conditions?

EXERCISE No. 11.4 — Voltage control of a shunt dynamo.

Apparatus

Shunt dynamo coupled to a constant speed motor (an a.c. induction motor is suitable for this purpose).
Field regulator.
Ammeter.
Voltmeter.

Diagram

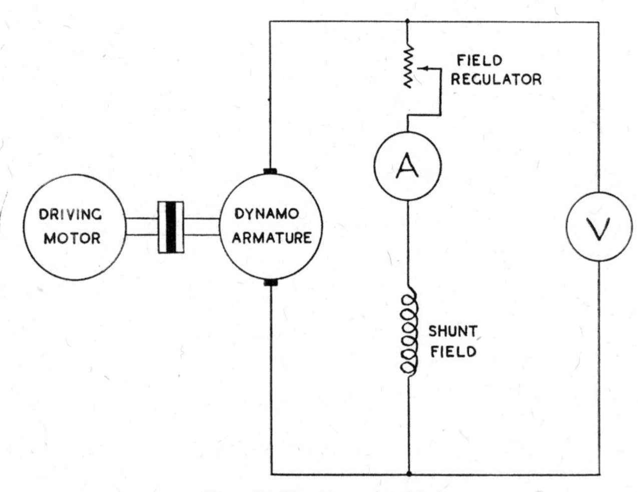

FIG. 11.20 Exercise 11.4

Procedure

1. Connect the circuit as shown in Fig. 11.20
2. Start the motor, adjust the field regulator to give maximum field current and note the generated voltage and field current.
3. Reduce the field current by a small amount again noting voltage and current.
4. Repeat step 3, reducing the field current in small steps to zero.
   Plot a graph of generated voltage on a base of field current.

Answer the following questions

1. Explain carefully the effect of connecting a rheostat in series with the field winding of a shunt dynamo.
2. Why is residual magnetism important for the successful operation of a self-excited dynamo?

CHAPTER 12

A. C. Motors

12.1 Production of Rotating Magnetic Fields

(*a*) A three-phase a.c. supply may be used to produce a rotating magnetic field. If three coils are situated 120° apart and connected to the RED, YELLOW and BLUE phases of a three-phase a.c. supply then the currents at various instants are shown by the wave diagrams in Fig. 12.1.

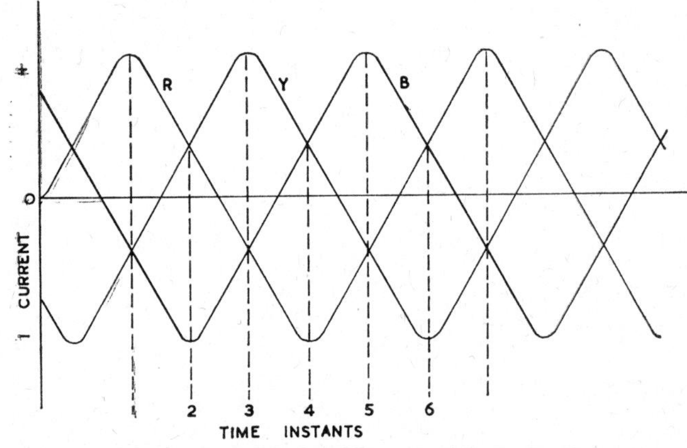

FIG. 12.1 Wave Diagram for Three-Phase Currents

(*b*) Consider the time-instant marked 1 on the wave diagram. The red phase coil is carrying maximum current in the positive direction so producing a peak magnetizing force in line with itself. At the same time-instant the blue and yellow phase coils are each carrying a negative current of half maximum value so producing a half-peak value magnetizing force in line with each coil. Fig. 12.2(a shows the three coils and the directions of the magnetizing forces at time-instant 1.

(*c*) These forces can be added vectorially as shown in Fig. 12.2(b), where OA represents the magnetizing force of the red phase coil in magnitude and direction; similarly AB represents the magnetizing force of the blue phase coil and BC represents the magnetizing force of the yellow phase coil. OC represents the resultant magnetizing force of the three coils acting together.

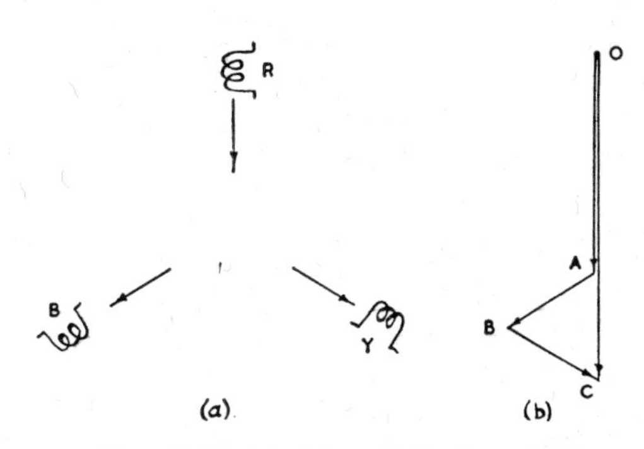

FIG. 12.2 (a–b) Magnetizing Force Vectors

(a) Directions of magnetizing forces; (b) Vector addition of magnetizing forces.

(*d*) By repeating the process the resultant magnetizing force for other instants of time can be found. Fig. 12.3 shows the direction of the resultant magnetizing force for the time instants marked 1 to 6 on the wave diagram.

(*e*) From the above it can be seen that a three-phase winding can produce a magnetic field which remains constant in strength but whose direction rotates at a constant speed. In practice a rotating magnetic field can be produced by a three-phase winding situated in slots in the inner circumference of the fixed part of the motor which is called the stator. The field produced may be a two-pole field, making one revolution per cycle, a four-pole field making one revolution in two cycles, and so on. Any even number of poles may be used, a 'p'-poled field making one revolution in $\frac{p}{2}$ cycles. Thus

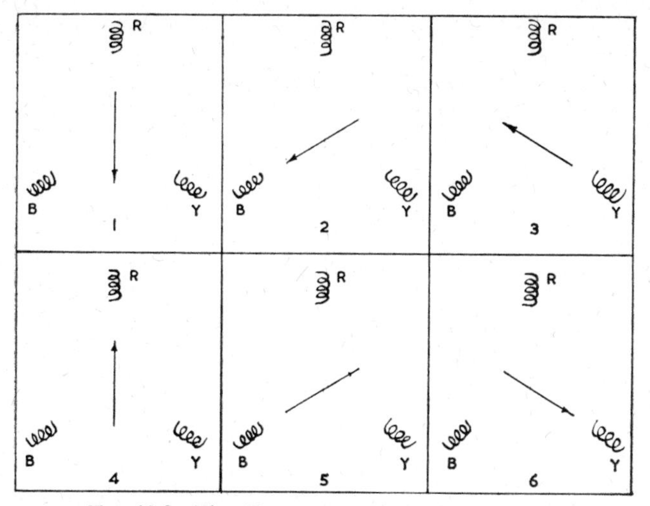

FIG. 12.3 Flux Vectors at Various Time Instants

the speed at which the field rotates is $\frac{2f}{p}$ revolutions per second, where f is the supply frequency. This speed is known as the synchronous speed and is more conveniently calculated from the formula:

$$N_S = \frac{60 \times 2f}{p} \text{ r.p.m.}$$

Two types of a.c. motor which use a rotating magnetic field are the synchronous motor and the induction motor.

12.2 Synchronous Motors

The rotating part or 'rotor' of a synchronous motor is magnetized with a fixed polarity. In small motors the rotor may be permanently magnetized while in larger motors the rotor carries a winding which is fed with direct current via two slip rings. This type of motor will run only at synchronous speed when the permanent field of the rotor 'locks' with the rotating magnetic field of the stator. Small synchronous motors are used in electric clocks because their speed is completely governed by the frequency of the main supply. Larger types of synchronous motor are sometimes used because they may be made to operate at a leading power factor, but their use is limited owing to the complicated starting equipment required.

12.3 Induction Motors

(*a*) The induction motor is the most widely used type of a.c. motor owing to its comparatively low cost and high efficiency.

Two types of rotor may be used:

(i) The 'squirrel cage' rotor.
(ii) The 'wound' rotor.

The winding of a squirrel cage rotor usually consists of copper or aluminium bars fitted in slots, the bars being connected to end rings at each end of the rotor. With this type of rotor the rotor bars are not connected directly to the source of supply so that the operation of the motor depends on currents induced in the rotor bars by the rotating magnetic field. In order to induce the necessary rotor currents, the magnetic field must rotate faster than the rotor, therefore the actual speed (N_r) of an induction motor is always slightly below the synchronous speed (N_S). The fractional slip of an induction motor is defined as:

$$\text{Fractional slip} = \frac{N_S - N_r}{N_S}$$

The squirrel cage induction motor is widely used as it is both cheap and efficient, having only one real disadvantage in that its starting torque is comparatively low. When a high starting torque is needed then a slip ring motor, using a wound rotor, may be used.

(*b*) The wound rotor has a winding made up of coils situated in slots. The coils are usually connected so that they form a three-phase 'star'-connected winding with the outer ends of the coils brought out to three slip rings. When the motor is running at its normal speed the slip rings are short-circuited by the motor starter and the motor operates in the same way as the squirrel cage motor. But, when starting, the rotor resistors are connected in series with the slip rings. This not only reduces the starting current but also increases the starting torque.

12.4 Three-phase Induction Motor Starters

(*a*) The principal types of starter used for three-phase induction motors are:

(i) Direct-on-line starter.
(ii) Star-delta starter.
(iii) Rotor resistance starter.
(iv) Auto-transformer starter.

The direct-on-line starter is used for the smaller type of motor where the value of starting current caused by applying full line voltage to the stationary motor is not excessive. The starter is usually in the form of a 'contactor', i.e. a magnetically-operated switch. When the 'start' button is pressed, current flows through the magnetizing coil of the contactor so attracting the armature and closing the contacts. This connects the motor to the supply and, at the same time, provides an auxiliary connection to the magnetizing coil so that the 'start' button may be released. In the event of a supply failure the magnetizing coil is de-energized so releasing the armature and opening all contacts. Thus, a contactor operated starter automatically gives NO-VOLTS protection. Overload protection may be afforded by either bi-metal strips with heat coils or by magnetically operated overload trips. In either case operation of the overload device opens a contact, so interrupting the current to the magnetizing coil. The 'stop' button is wired in series with the magnetizing coil; when the button is pressed the circuit to the magnetizing coil is broken so releasing the armature and switching off the motor. Fig. 12.4 shows a typical direct-on-line starter. When the 'ON' button is pressed a circuit is made from phase 2, through the 'ON' and 'OFF' switches and the tripping switch, T, to coil, C, and on to phase 3. The coil becomes energized and attracts the moving contacts so that the supply is switched on to the motor. Terminal 5 is now connected to phase 2 so that the coil, C, will remain alive when the 'ON' button is released. When the 'OFF' button is pressed the coil circuit is broken so releasing the contacts. In the event of an overload one or more of the heat coils will become hot, causing a bi-metal strip to bend and operate the tripping switch, T, once again breaking the coil circuit and releasing the contacts.

(*b*) The star-delta starter is used for medium size motors, its purpose being to reduce the current at starting. The stator winding is divided into three sections, one for each phase. The starter consists of a changeover switch having three positions, 'off', 'start', and 'run', and is spring-loaded to return to the 'off' position. In the 'off' position no connection is made to the motor. In the 'start'

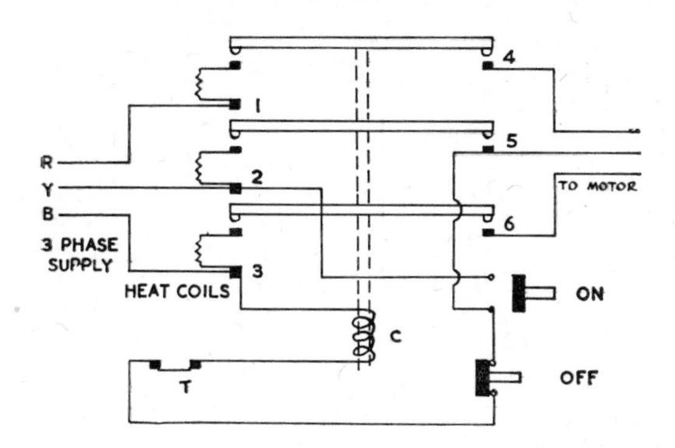

FIG. 12.4 Direct-on-line Contactor Starter

position the stator winding is star-connected and so receives $\frac{1}{\sqrt{3}}$ of the line voltage. In the 'run' position the stator winding is delta-connected and so receives full line voltage. A mechanical interlock is normally incorporated to prevent the starter being immediately switched to 'run' so that when starting the motor, it is always necessary to select 'start' first. When in the 'run' position the starter-switch is held on by a catch which can be released by the NO-VOLTS coil or the overload trips; when the catch is released the starter automatically returns to the 'off' position. Fig. 12.5 shows the action of a typical star-delta starter switch.

(*c*) The rotor resistance starter is used in connection with a slip-ring motor when a high-starting torque is required. The stator circuit is connected by a direct-on contactor starter while the rotor circuit is completed via the slip-rings and starting resistors. As the motor gathers speed the starting resistors are progressively cut out of circuit until finally the slip-rings are short-circuited. Electrical interlocking is normally incorporated in the starter to ensure that the rotor resistors are correctly connected for starting before the stator contactor can be operated. Fig. 12.6 shows the principle of the rotor resistance starter.

(*d*) The auto-transformer starter uses a three-phase auto-transformer to reduce the voltage applied to the motor when starting and so reduces the starting current. By switching to successive tappings the voltage applied to the motor is progressively increased

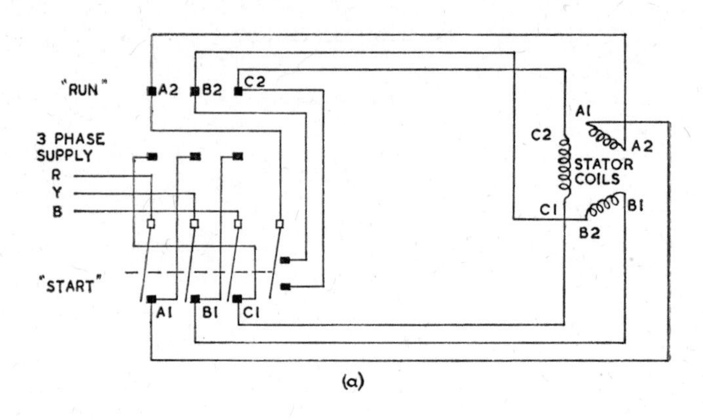

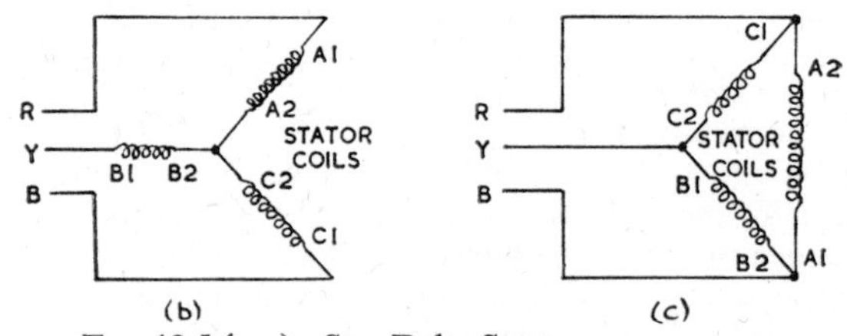

FIG. 12.5 (a–c) Star-Delta Starter
(a) Circuit diagram; (b) Starting connections; (c) Running connection.

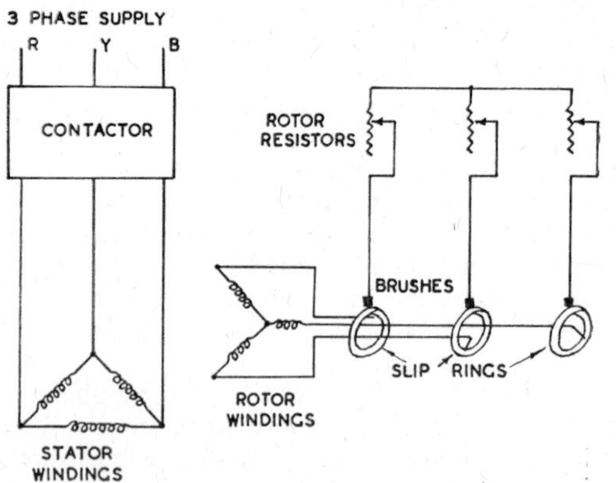

FIG. 12.6 Rotor Resistance Starter

until, when the motor is up to speed, the auto-transformer is cut out of circuit and the motor receives the full supply voltage. Fig. 12.7 shows the principle of the auto-transformer starter.

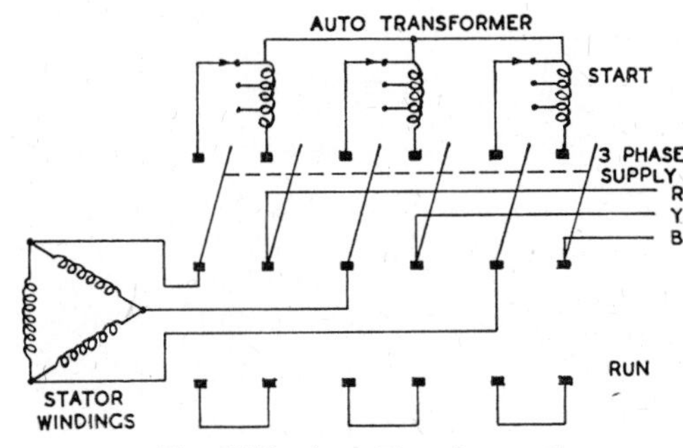

FIG. 12.7 Auto-Transformer Starter

12.5 Single-phase Induction Motors

(*a*) A single-phase a.c. supply produces a pulsating rather than a rotating magnetic field, nevertheless single-phase a.c. induction motors will operate successfully provided that an initial start is given to the rotor. The single-phase motor will run equally well in either direction, so that the direction in which it rotates is determined by the direction of the initial starting torque. Two common methods of providing the necessary starting torque are:

(i) Split-phase start.
(ii) Capacitor start.

(*b*) Motors intended for 'split-phase' starting have a 'running' winding and a 'starting' winding; the starting winding possesses a high inductance so that the current flowing through it lags the supply voltage by a large phase angle. This has an effect equivalent to a two-phase supply and so produces a rotating magnetic field which starts the motor. When the motor has gained sufficient speed a centrifugal switch disconnects the starting winding. Fig. 12.8 shows the connections for a split-phase start motor.

(*c*) Capacitor start motors also have two windings, but in this case the starting winding has a comparatively low inductance and is connected in series with a capacitor. The series capacitor causes

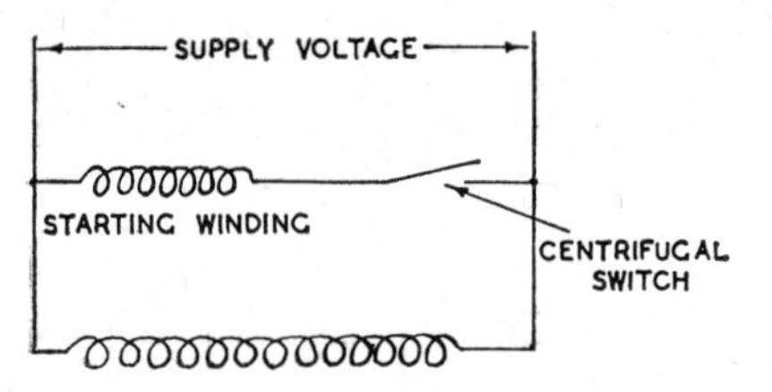

FIG. 12.8 Split-Phase Start Motor

the current flowing through the starting winding to lead the supply voltage so, once again, the effect of a two-phase supply is obtained, resulting in the production of a rotating magnetic field which starts the motor. When the motor is up to speed the starting winding is usually disconnected automatically by a centrifugal switch; but in some cases the starting winding may continue to receive current via a smaller capacitor. Fig. 12.9 shows the connections of a capacitor start motor.

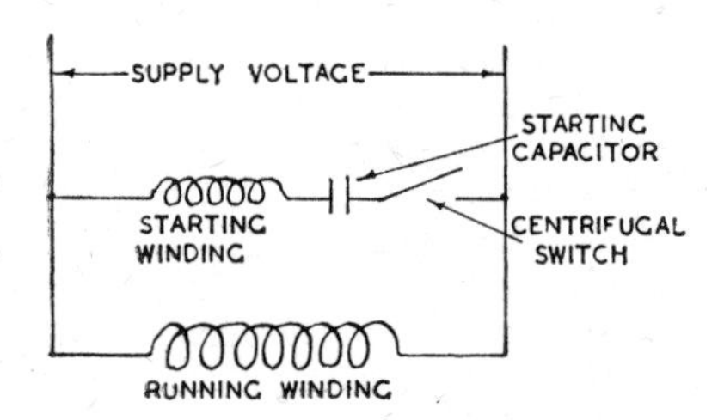

FIG. 12.9 Capacitor Start Motor

12.6 Reversal of Induction Motors

The direction of rotation of a three-phase induction motor may be reversed by interchanging the connections of any two of the three incoming lines. Since the direction of rotation of a single-phase induction motor depends on the direction in which it is started, single-phase induction motors may be reversed by reversing the connections to the starting winding.

EXERCISES

The exercises which conclude this chapter are designed mainly to demonstrate the principal methods of starting induction motors.

EXERCISE No. 12.1 — To connect a three-phase induction motor using a direct-on-line contactor starter.

Apparatus

Three-phase induction motor.
Direct-on-line contactor starter.

Procedure

1. Examine the terminal blocks in the motor and starter and the connection diagram inside the lid of the starter.
2. Connect the motor to the starter and the starter to the isolator in accordance with the connection diagram.
3. Switch on and start the motor, noting the direction of rotation.
4. Switch off and reverse any two of the incoming lines at the starter terminals.
5. Switch on and start the motor again, noting the direction of rotation. Make a neat circuit diagram of the motor and starter circuit.

Answer the following questions

1. Why is this type of starter not usually used to start the larger types of induction motor?
2. How can the direction of rotation of three-phase induction motors be reversed?

EXERCISE No. 12.2 — To connect a three-phase induction motor using a star-delta starter.

Apparatus

Three-phase induction motor.
Star-delta starter.
500V range a.c. voltmeter.

Procedure

1. Examine the terminal blocks in the motor and starter and the connection diagram inside the lid of the starter.
2. Connect the motor to the starter and the starter to the isolator, in accordance with the connection diagram.

3. Switch the motor on in accordance with the directions on the starter, and note the direction of rotation.
4. Switch off and reverse any two of the incoming lines at the starter terminals.
5. Switch on, and start the motor and note direction of rotation.
6. Switch off and connect the voltmeter to terminals A1 and A2 at the motor terminal block.
7. Switch on and start the motor, noting the voltmeter reading

   (*a*) when the starter-switch is in the 'START' position
   (*b*) when the starter-switch is in the 'RUN' position.

   Make a neat circuit diagram of the motor and starter circuit.

Answer the following questions

1. What are the relationships between 'line' and 'phase' voltage:

   (*a*) For STAR connections.
   (*b*) For DELTA connections.

2. What were the actual values of voltage applied to the motor-phase coil as measured in step 7 above? Do these values agree with the theoretical relationship for STAR and DELTA connections?

EXERCISE No. 12.3 — To connect a three-phase induction motor using an auto-transformer starter.

Apparatus

Three-phase induction motor.
Auto-transformer starter.
500V range a.c. voltmeter.

Procedure

1. Examine the terminal blocks in the motor and starter and the connection diagram inside the lid of the starter.
2. Connect the motor to the starter and the starter to the isolator in accordance with the connection diagram.
3. Switch the motor on in accordance with the directions on the starter.
4. Switch off and connect the voltmeter to terminals at the motor terminal block.
5. Switch on and operate the starter slowly, noting the voltmeter reading at each position of the starter handle.
   Make a neat circuit diagram of the motor and starter circuit.

Answer the following questions

1. What is the function of the auto-transformer in the starter?
2. What are the requirements of the I.E.E. Regulations concerning the use of auto-transformers?

EXERCISE No. 12.4. To connect a three-phase slip-ring induction motor using a rotor resistance starter.

Apparatus

Three-phase slip-ring motor.
Rotor resistance starter.

Procedure

1. Examine the terminal blocks in the motor and starter and the connection diagram inside the lid of the starter.
2. Connect the motor to the starter and the starter to the isolator in accordance with connection diagram.
3. Switch the motor on in accordance with the directions on the starter.
   Make a neat circuit diagram of the motor and starter circuit.

Answer the following questions

1. For what type of loading is the slip-ring induction motor best suited?
2. What are the requirements of the I.E.E. Regulations concerning the size of cables used for connecting the rotor circuits of slip-ring motors?

CHAPTER 13

Installation and Maintenance of Motors and Generators

When choosing a motor for a particular installation the following points have to be taken into consideration:

(*a*) The type of supply available.
(*b*) The nature of the load.
(*c*) Any special adverse conditions such as excessive heat, moisture, explosive or inflammable atmospheres.

13.1 Motors for use on d.c. Supplies

(*a*) The d.c. series motor provides a high starting torque but is liable to race if disconnected from its load. This makes the motor suitable for direct-coupled loads such as fans, and for traction purposes but this type of motor is NOT suitable for belt drives.

(*b*) The d.c. shunt motor provides a reasonable starting torque and a fairly constant speed, and the speed can be easily controlled by means of a field regulator. Its characteristics make the shunt motor suitable for most general purpose drives.

(*c*) The cumulative compound motor can provide a high starting torque together with a safe no-load speed; it is particularly suitable for heavy intermittent loads such as fly presses. It is also suitable for loads requiring a high starting torque and a fairly constant speed drive such as compressors and hoists. A further application is for loads possessing a high inertia and subject to sudden overloads, such as rolling mills.

13.2 Motors for use on Three-phase a.c. Supplies

(*a*) The squirrel cage induction motor is the least expensive type of motor and is suitable for most general-purpose drives, its only real disadvantages being that the starting torque is somewhat limited and the speed cannot easily be varied. Double, and triple, cage rotor machines are available for use where a high starting torque is important.

(*b*) The slip-ring induction motor provides a high starting torque but is more expensive than the squirrel cage type.

(*c*) The synchronous motor can be used to provide an absolutely constant speed drive. A feature of this motor is that it can be designed to operate with a leading power factor, so providing power factor correction for other equipment in the installation.

(*d*) There are various types of variable speed a.c. motor, e.g. the Commutator motor. These types of motor are comparatively expensive and so they are only used where variable speed operation is essential.

13.3 Motors for use on Single-phase a.c. Supplies

Single-phase induction motors can be used to provide constant speed drives for the smaller loads. As they are not so efficient as three-phase motors they are seldom recommended for use with loads requiring much more than one horse-power. Series connected (universal) motors can be used on either a.c. or d.c. supplies; they are commonly used in vacuum cleaners, small electric drills, etc.

13.4 Types of Motor and Generator Enclosure

(*a*) It is desirable that the outer casing of a motor or generator shall completely enclose and protect the windings and moving parts, both to prevent the ingress of dirt or moisture, and to protect the user from possible danger from the moving parts or live connections. At the same time the type of enclosure or casing used must allow the heat produced in the machine during normal operation to be safely dissipated, so preventing an excessive temperature rise in the windings. The types of enclosure used in practice are a compromise between the conflicting requirements of protecting the internals of the machine and providing sufficient ventilation to carry off the heat. The principal types of enclosure in general use are:

(i) Open type.
(ii) Screen-protected.
(iii) Drip-proof.
(iv) Totally enclosed.
(v) Flame-proof.

(*b*) Open type machines have no enclosure other than that provided by the yoke and bearing supports; there is thus no impediment to the ventilation and the cost of the machine is a minimum. As there is virtually no protection at all for the windings or moving parts, this type of machine should be installed only in clean, dry situations, where it is under the charge of skilled personnel, handrails being provided as necessary to prevent persons from accidentally touching the moving parts. Open type enclosures are usually restricted to the larger machines installed in generating stations and similar situations.

(*c*) The screen-protected type of machine uses perforated metal screens to enclose the ends of the machine. While this gives protection against inadvertent contact with the moving parts and also prevents large objects from falling into the motor it does not interfere unduly with the ventilation. Screen protection does not provide much protection against the ingress of dirt or moisture so this type of enclosure is recommended for general use in dry and reasonably clean situations.

(*d*) The drip-proof type of machine has its ends completely enclosed except for narrow slots or louvres which allow ventilation. These slots are usually situated on the underside of the machine and in some cases are backed by wire mesh screens. A fan is often incorporated to increase the circulation of air through the machine. These machines can withstand a small amount of water dripping on them but the degree of protection is not sufficient to withstand very wet or dirty conditions.

(*e*) A totally enclosed machine provides adequate protection to the internal parts in situations where very damp or dirty conditions prevail. This type of machine is often larger and more expensive than its screen-protected or drip-proof counterpart, owing to the need to dissipate all the heat produced internally from the outer surface of the casing.

(*f*) Flame-proof machines are used in situations where fire hazards due to inflammable or explosive atmospheres are present. These machines are totally enclosed and the lengths and clearances of the bearings, the design of the flanges and type of terminal box employed must all comply with B.S.S.229. When installing or maintaining this type of machine great care must be taken to ensure that any covers removed are replaced correctly and all fixing bolts and nuts replaced and correctly tightened, because the slightest error can destroy the flame-proof properties of the machine.

13.5 Installation of Motors and Generators

(*a*) It is normally essential that motors and generators are securely fixed in position so that their shafts will remain in correct alignment with the shafts of other equipment coupled to them. In the case of many machine-tools a mounting-plate is provided to which the motor can be fixed.

(*b*) Motors up to about 15 h.p. can be mounted by bolting on to slide-rails or to steel channels these, in turn, being bolted to structural steel work. Alternatively the motor feet, or the slide-rails, can be fixed using foundation bolts grouted into a concrete floor or by using expanding bolts. Thin steel shims may be placed under the motor feet near the fixing bolts to adjust the final alignment of the motor. The bedplates of small motor generator sets may be dealt with in the same way.

(*c*) Larger machines may require a specially prepared concrete base of sufficient thickness and strength to support the weight of the machine, and to withstand the pull of the belt-drive if this method of coupling is used. Concrete foundations should be at least 6 in. longer and wider than the fixing feet or bedplate of the machine and the sides should taper outwards slightly (about 2 in. per ft.). The base should be laid if possible on well rammed solid ground, allowing approximately 6 in. below ground for a 10-h.p. motor, increasing to about 2 ft. below ground for a 100-h.p. machine. If the site is not sufficiently firm, the area and thickness of the base may need to be increased to give adequate support for the machine.

13.6 Methods of Coupling

(*a*) Various means may be used to couple a motor to its load or a generator to its prime mover. The coupling can be provided by means of a belt running over pulleys fitted to the motor and load. An advantage of this method of drive is that the motor speed can be matched to that of the load by using pulleys of differing diameters:

$$\text{Load Speed} = \text{Motor Speed} \times \frac{\text{Diameter of Motor Pulley}}{\text{Diameter of Load Pulley}}$$

In some cases the ratio of the required load speed to the motor speed is so great that either an excessively large or small diameter pulley would be required to effect the speed change in one stage. In such cases the motor may be coupled to an ‘intermediate’ shaft which is in turn coupled to the load, so enabling the speed to be changed in two steps.

(*b*) The following points should be observed when using flat driving belts:

(i) The distance between the pulley centres should not be less than about 10 ft.
(ii) The pulleys must be in line, and the shafts parallel.
(iii) The pulleys should not be vertically above each other.
(iv) Adequate guards should be provided to prevent persons coming into contact with moving belts.
(v) If possible it should be arranged that the 'slack' side of the belt is uppermost.

Fig. 13.1 shows the arrangement of a flat belt-drive.

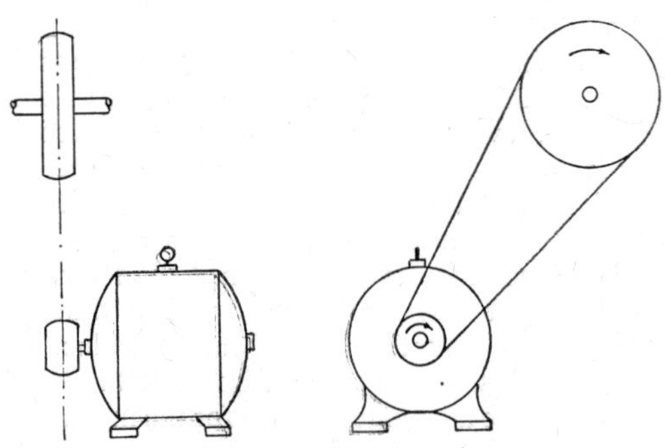

FIG. 13.1 Belt Drive

(*c*) Vee belts (see Fig. 13.2) are used when there is only a small distance between the shaft centres. This type of belt is manufactured in particular sizes, thus it is essential that the correct size of belt is ordered for any particular drive. The driving motor must be fitted

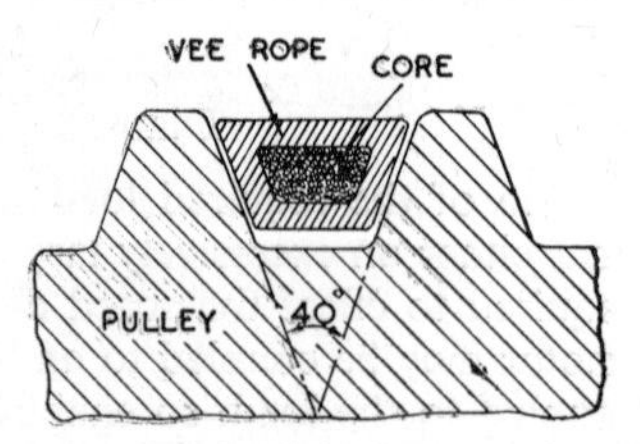

FIG. 13.2 Vee Belt and Grooved Pulley

on slide-rails or a similar device so that the belt can be slackened for fitting and removal and can be tensioned as required to drive without slipping. It is important to avoid over-tensioning the belt as this can cause excessive wear in the bearings and in some cases overheating of the motor.

(*d*) Multiple rope-drives (see Fig. 13.3) are suitable for transmitting large powers. The main advantages are:

(i) They are quiet in operation.
(ii) If one rope fails it is possible to carry on with only the remaining ropes until it is convenient to effect a replacement.
(iii) The pulleys can be slightly out of line without serious ill-effects.

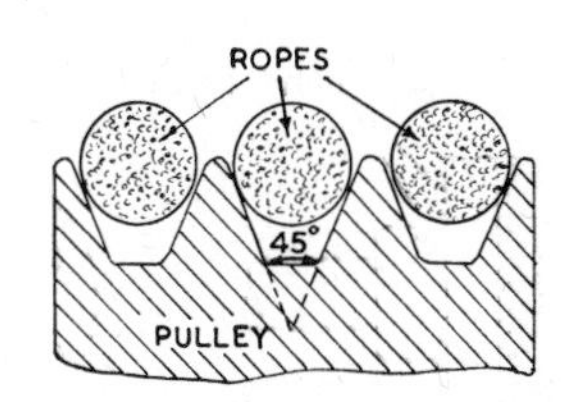

FIG. 13.3 Driving Ropes and Grooved Pulley

(*e*) The chain-drive provides an alternative to the belt-drive, its advantages being that the drive is positive and cannot slip. A disadvantage is that some of these drives are inclined to be noisy. It is essential that chain-drives are efficiently lubricated and well guarded. A good system is that where the chain is totally enclosed and runs through an oil-bath.

(*f*) Solid couplings can be used only where it is possible to maintain the driving and driven shafts in perfect alignment, for example, where a generator and its motor are mounted on the same bed-plate. Fig. 13.4 illustrates some common faults of alignment. It cannot be too strongly emphasized that correct alignment of the shafts is essential where this type of coupling is employed.

(*g*) Many types of flexible coupling are available, one example being shown in Fig. 13.5. The advantages of flexible couplings are:

(i) They prevent vibration being transmitted along the shafts.
(ii) They help to absorb the impact of sudden loads applied to the shafts.

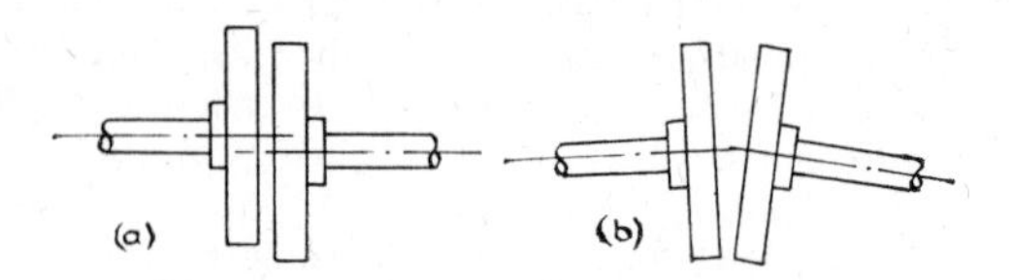

FIG. 13.4 (a–b) Faults in Aligning Solid Couplings
(a) Shafts not in line; (b) Shafts not parallel.

(iii) They will operate successfully with the shafts slightly out of alignment, although with most types it is still advisable to align the shafts as accurately as possible.

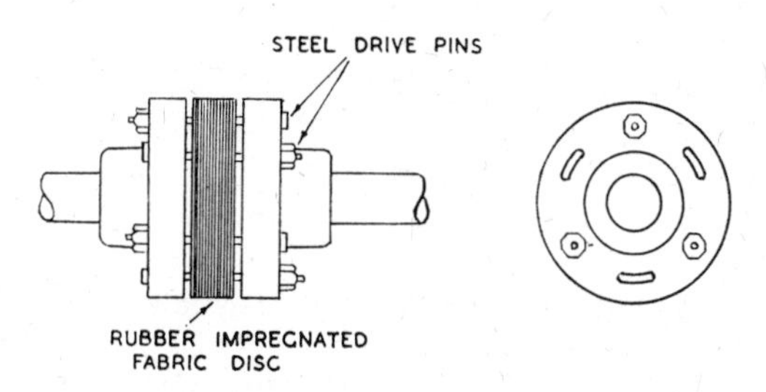

FIG. 13.5 Flexible Coupling

13.7 Wiring Requirements for Motors

(*a*) Recommendations for good practice when installing electric motors are given in Code of Practice C.P.321, 102. It is also necessary to comply with the requirements of the I.E.E. Wiring Regulations and when installing motors in factories the provisions of the Factories Act must be observed. Some of the more important points which must be observed are discussed below.

(*b*) Every electric motor must be provided with efficient means of starting and stopping. The means of stopping the motor must be within easy reach of the operator of the machine driven by the motor, and in many cases this means that remote 'stop' buttons must be fitted.

(*c*) Every motor exceeding $\frac{1}{2}$ horse-power rating should be provided with:

(i) Efficient means of isolating the motor and its control gear from the supply. If the isolator is remote from the machine then provision must be made for it to be locked in the 'off' position.

(ii) Protection against excess current. Normally the rating of the fuses used to protect the final sub-circuit (or one half the setting of the circuit-breaker) must equal the rating of the cables used but, if the motor starter itself provides excess current protection for the motor and the cables between motor and starter, then the fuse rating (or one-half the setting of the circuit-breaker) may be increased up to twice the rating of the cables used, in order to accommodate the starting current of the motor.

(iii) No-volt release, to prevent restarting after failure of supply. (This requirement is relaxed in special circumstances such as in cases where failure of the motor to restart as soon as the supply is restored gives rise to dangerous conditions, or where the motor normally starts at irregular intervals under automatic control.)

(*d*) The frame and all-exposed non-conducting metal parts must be efficiently earthed. It should be noted that it is often very convenient to employ flexible metallic conduit to protect the final connections to the motor. This does not in itself provide efficient earth continuity and a separate earth continuity conductor must always be used.

(*e*) The supply cables must be rated to carry at least the full load current of the motor. In some cases it may be advisable to increase the cable sizes above the minimum in order to reduce the voltage drop when the motor is starting. It should be noted when selecting cable sizes for the rotor circuits of slip-ring induction motors that the rotor current may be several times greater than the supply current and that it is important to keep the resistance of the rotor circuits to the minimum practical value if the motor is to operate efficiently.

13.8 Routine Maintenance

(*a*) The purpose of routine maintenance is to ensure, as far as possible, that a machine is operating, and will continue to operate, in a safe and efficient manner. A systematic inspection routine can result in minor faults being detected and corrected before serious trouble occurs. The frequency with which plant should be inspected is determined by the conditions under which it operates, and the danger and expense which may arise due to a breakdown. The tasks which are normally carried out during a routine inspection are discussed briefly below:

(i) Any deposits of dust and dirt, especially from around brush holders and from ventilation openings, should be removed.
(ii) All fixing bolts and driving arrangements should be checked for security, and it should be ensured that all cover plates and guards are correctly in place.
(iii) Bearings should be checked for excessive play and correct lubrication; excessive lubrication must be avoided owing to the risk of oil or grease finding its way on to the windings.
(iv) Brushes must be free to slide in their holders; the tension should be sufficient to hold the brush firmly against the commutator or slip-ring. Badly worn brushes should be replaced. As a commutator forms a skin during service excessive cleaning will do more harm than good, but if the commutator is badly blackened it may be cleaned by wiping with a rag moistened with paraffin. The grooves between the commutator segments should be kept free from carbon dust, etc., as this may lead to short-circuiting of the armature coils and excessive sparking.
(v) An overall insulation test should be carried out between the windings and the frame of the machine preferably including the starter and all the motor wiring from the isolator.
(vi) The effectiveness of the earthing should be checked, this test being particularly important in the case of portable tools.
(vii) Starters and control gear should be checked for burnt contacts and for correct operation of protective devices. In particular it should be confirmed that remote 'stop' buttons, and any interlocking devices, are effective.

(*b*) It is good practice to keep a record of the work carried out, and the readings of insulation and earth continuity resistance obtained during routine maintenance. A note should be made of the atmospheric conditions at the time of testing since this may affect the insulation resistance. The intelligent use of records can often lead to the detection of a fault before it becomes serious; for example, an unexpectedly high reading for earth continuity, although still within the acceptably safe limit, may point to such troubles as fractured strands in the earth continuity conductor, or loose or corroded earthing terminals. If no past records were available the reading might have been accepted without further investigation possibly allowing a dangerous situation to arise.

13.9 Fault-finding

(*a*) The location and nature of a fault in an electrical machine can usually be quickly determined by systematic tests. The tests to be carried out obviously depend upon the symptoms of the fault; for example, if a motor shows no sign of life at all it is advisable to check the supply first. Most of the tests which are required can be carried out using:

(i) Voltage indicator, test lamp, or voltmeter for checking the supply.
(ii) Continuity tester.
(iii) Insulation tester.

(*b*) It is obviously impossible in a book of this size to list every type of fault which may occur; the table below lists some typical fault symptoms and the tests which may be employed to determine the nature of the fault.

Symptoms	*Tests*
Motor completely dead.	Check voltage at isolator and motor terminals.
Contactor starter does not operate, although supply at isolator is correct.	(i) Check that overload trips, limit trips, interlocks and remote stop buttons are not operated, and that starter controls are correctly set to start position. (ii) Test continuity of contactor coil and its associated circuits.
Fuses blow or overload trips operate when any attempt is made to start the motor.	(i) Check that motor is free to rotate. (ii) Check that starter is being operated correctly. (iii) Test insulation resistance.
Three-phase motor buzzes or hums but refuses to start.	(i) Check that supply voltage is available at *all* three phases at motor terminals. (ii) Test each phase of motor winding for continuity. (iii) Test rotor circuits for continuity.
Single-phase motor hums but refuses to start.	(i) Check that motor is free to rotate (particularly for small-sized motors). (ii) Test continuity of main and starting windings and of centrifugal switch. (iii) Test that supply is actually reaching the starting winding via the capacitor if fitted.

Symptoms	*Tests*
Starter of d.c. motor will not hold in 'ON' position although motor starts correctly.	(i) Check that overload trip is not stuck in operated position, and that remote 'stop' buttons, etc., are not operated. (ii) Test 'no-volts-coil' for short-circuit. (iii) In the case of series motors, or starters where no-volts coil is *not* in series with a shunt winding, check the no-volts coil circuit for continuity.
Excessive sparking at commutator.	(i) Check brushgear for correct tension, brushes sticking in holders, etc. (ii) Check polarity of interpoles if fitted. (iii) Test armature windings.
D.C. dynamo fails to excite.	(i) May be due to loss of residual magnetism; 'flash' using a suitable battery. (ii) Check that rotation of dynamo is in correct direction, and that field connections have not been reversed. (iii) Check field and armature circuits for open or short circuits.

EXERCISES

The exercises that conclude this chapter are designed to give practice in the maintenance of electrical machines.

EXERCISE No. 13.1 — Routine inspection and tests of a d.c. shunt motor.

Apparatus

D.C. Shunt motor.
Face-plate starter.
Field regulator.
(These three should be already connected.)
Insulation tester.

Procedure

1. Ensure that the supply is OFF.
2. Carry out a visual inspection of the motor, paying particular attention to the following points:

   (*a*) Presence of oil or grease on windings.
   (*b*) Obvious mechanical damage.
   (*c*) Condition of bearings.
   (*d*) Condition of commutator and brush gear.

3. Inspect the starter, paying particular attention to the following points:

   (*a*) Condition of contact studs.
   (*b*) Condition of no-volt and overload protection devices.

4. Inspect the field regulator, paying particular attention to the following points:

   (*a*) Condition of contact studs.
   (*b*) That the regulator control handle moves smoothly.

5. Carry out necessary insulation tests.
6. Switch on supply and check that the motor and its control gear function correctly.
7. Prepare a comprehensive report on the condition of the motor including results of insulation tests, noting weather conditions.

EXERCISE No. 13.2 Routine inspection and tests of a three-phase squirrel cage induction motor.

Apparatus

Three-phase squirrel cage induction motor.
Star-delta starter.
} These should be already connected.
Insulation tester.

Procedure

1. Ensure that the supply is OFF.
2. Carry out a visual inspection of the motor, paying particular attention to the following points:
   (*a*) Presence of oil or grease on windings.
   (*b*) Obvious mechanical damage.
   (*c*) Condition of bearings.
3. Inspect the starter, paying particular attention to the following points:
   (*a*) Condition of all contacts.
   (*b*) Condition of no-volts and overload protection devices.
4. Carry out necessary insulation tests.
5. Switch on supply and check that the motor and its control gear function correctly.
6. Prepare a comprehensive report on the condition of the motor including results of insulation tests, noting weather conditions.

EXERCISE No. 13.3 **Routine maintenance of a portable electric hand tool.**

Apparatus

Portable electric drill or similar apparatus.
Insulation tester.
High-current continuity tester.

Procedure

1. Inspect the tool and flexible connecting lead for signs of obvious damage.
2. Remove any accumulations of dirt, particularly from ventilation holes, around brush-holders, etc.
3. If the motor is fitted with brushes, check that they can slide freely in their holders and that they are not excessively worn. Inspect the commutator for signs of excessive wear.
4. Lubricate bearings as required, taking particular care not to allow any lubricant to get on to the electrical windings, commutator, etc.
5. Check that all fixing screws are tight, not forgetting the clamping screws for the flexible conductors. Replace any covers which have been removed for the above tests.
6. Check the insulation resistance between the 'live' pin of the plug top fitted to the flexible connecting lead and the frame of the tool.
7. Test for effective earth continuity between the earth pin of the plug top and the frame of the tool, using a type of tester which injects a current of approximately 25A.
8. If all the above tests are satisfactory connect the tool to the supply and check that it operates correctly.

   Prepare a detailed written report on the condition of the tool.

EXERCISE No. 13.4 — Faults on a d.c. generator.

A complaint has been received that a d.c. generator is not functioning. The object of this exercise is to locate the fault systematically and rectify it.

Apparatus

D.C. generator and driving motor (already connected).
Insulation and continuity tester.
Multi-range test meter.

Procedure

1. Start the motor and run the generator up to speed.
2. If the generator fails to generate, stop the machines, and carry out tests to determine the cause of the trouble.
3. Carry out the necessary repairs.
4. Write a complete report describing the symptoms and the steps taken to locate and rectify the fault(s).
5. Draw a complete diagram of the generator connections, showing the positions of the fault(s).

Note on Exercises 13.4 and *13.5*

Before the above exercises can be performed by the student it is necessary for the teacher to introduce suitable faults. Suggested faults are:

(a) *Induction Motor*
Open-circuited phase connection.
Open-circuited rotor connection.
Open-circuited operating coil in contactor starter, etc.

(b) *D.C. Generator*
Open-circuited field or armature connection.
Rotation reversed, short-circuited armature, etc.

EXERCISE No. 13.5 — Faults on a three-phase induction motor.

A complaint has been received that a three-phase induction motor is not operating satisfactorily. The object of this exercise is to locate the fault systematically, and rectify it.

Apparatus

1 Three-phase induction motor.
1 Rotor resistor starter.
Insulation and continuity tester.
Multi-range test meter.

Procedure

1. Try to start the motor in accordance with the manufacturer's instructions.
2. If the motor does not start, determine the cause or causes of the trouble and carry out the necessary repairs.
3. Write a complete report describing the symptoms and the steps taken in locating and rectifying the fault(s).
4. Draw a complete diagram of the motor and starter connections, showing the position of the fault(s).

CHAPTER 14

The College Workshop

14.1

The purpose of a college workshop is to provide facilities for the student to study the practical aspects of his chosen craft. Many electrical apprentices are employed by the smaller firms, and there is a danger that their working experience may be confined to only a limited field of electrical installation work. Thus it becomes particularly important that the college workshop is equipped to enable the student to carry out the widest possible variety of tasks, so widening his experience of electrical equipment and circuits which is essential if he is to become a fully proficient craftsman.

14.2 Safety Precautions

(*a*) It is inevitable that electrical installation workshops use supplies at normal mains voltages, and this obviously introduces an element of danger if safety precautions are not strictly observed.

The following two rules provide a basis for safe working in any teaching workshop:

(i) No student should be allowed to connect to, or switch on, any supply without the express permission of the teacher in charge.

(ii) All wiring exercises should be inspected, and tested using the appropriate instruments, before the student is permitted to connect them to a supply.

(*b*) It is an advantage if the main power supply to the workshop is controlled by a contactor. This contactor should be provided with remote 'stop' buttons situated in strategic positions in the workshop, so that the supply can be rapidly switched off should an emergency arise.

14.3 Demonstration Boards

(*a*) Demonstration boards can be used with advantage in the workshop for three main purposes:

(i) To illustrate the various steps of a particular operation.
(ii) To provide a convenient assembly of apparatus for carrying out particular exercises.
(iii) To simulate, within the confines of the workshop, conditions which occur in full-sized installations.

Examples of the use of such boards have been given in previous chapters. This chapter is concluded by the constructional details of the boards used.

Demonstration Board No. 1. Steps in Terminating P.I.L.C. Cable The cable ends, in various stages of preparation, and a terminal box, fitted with a perspex front cover, are mounted as shown in Fig. 14.1.

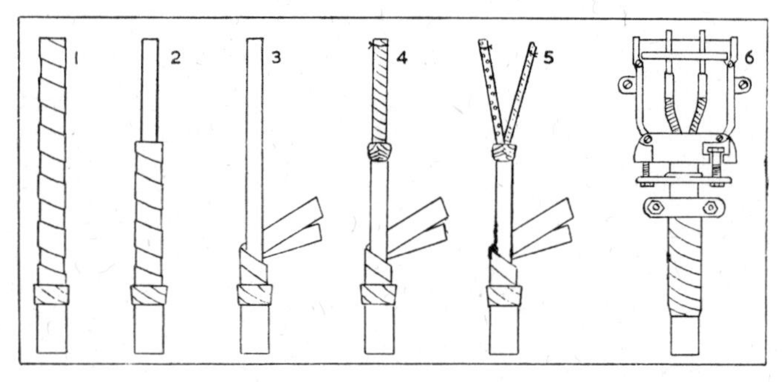

FIG. 14.1 Demonstration Board No. 1

Demonstration Board No. 2. Fuse and Miniature Circuit-Breaker. The M.C.B., fuse unit, two-way switch and socket outlet are arranged as shown in Fig. 14.2. The apparatus is connected so that the load connected to the socket outlet is supplied via either the M.C.B. or the fuse, according to the position of the two-way switch.

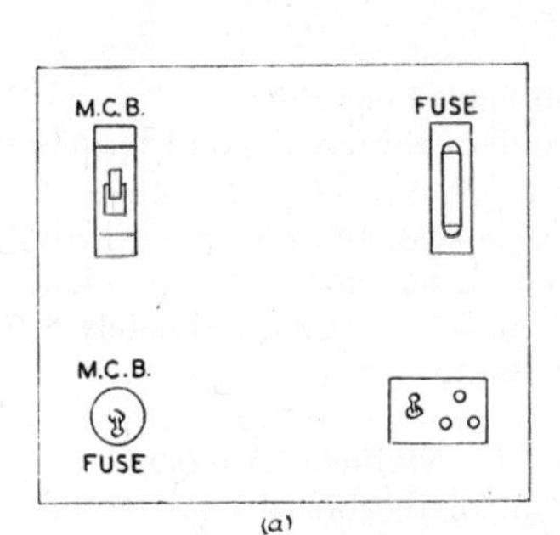

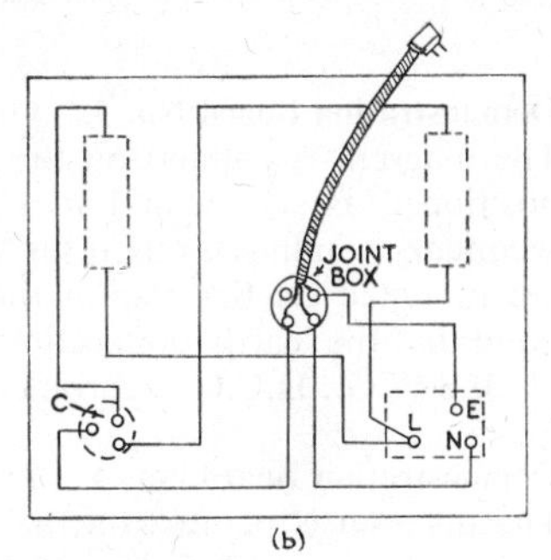

Fig. 14.2 (a–b) Demonstration Board No. 2
(a) Front View; (b) Rear view.

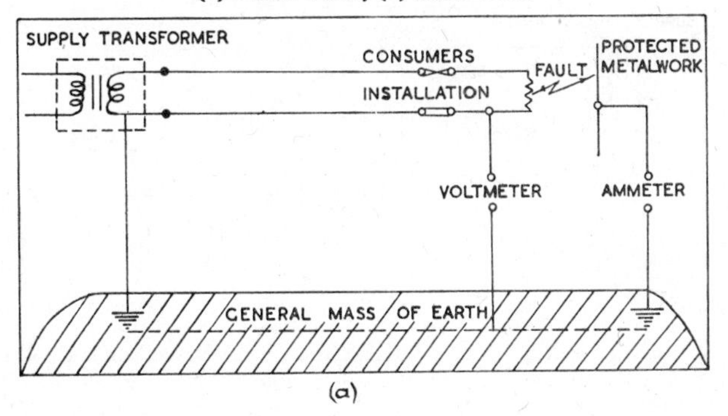

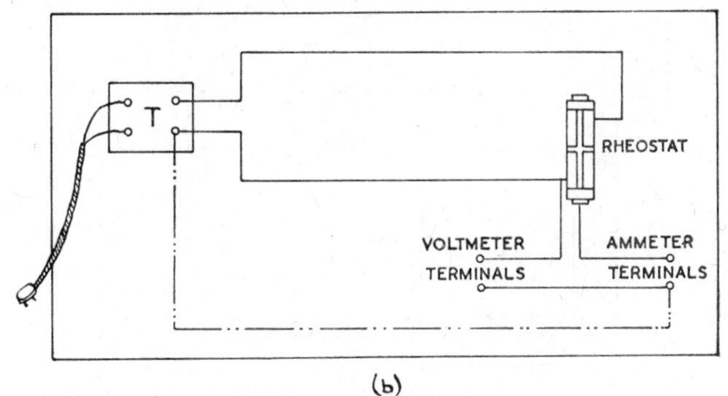

Fig. 14.3 (a–b) Demonstration Board No. 3
(a) Front view; (b) Rear view.

Demonstration Board No. 3. Earth Fault Loop Path
The diagram is painted on the front of the board, and terminals positioned as shown in Fig. 14.3(a). A transformer with a 12V secondary winding (as used for 'ray' boxes), and a 12-ohm rheostat are mounted on the rear of the board as shown in Fig. 14.3(b). Note that the 'earth' connection is made using approximately 5 ft. of 20 S.W.G., D.C.C. nichrome resistance wire.

Demonstration Board No. 4. Protective Multiple Earthing.
The diagram is painted on the front of the board, and terminals positioned as shown in Fig. 14.4(a). A transformer with a 12V sec-

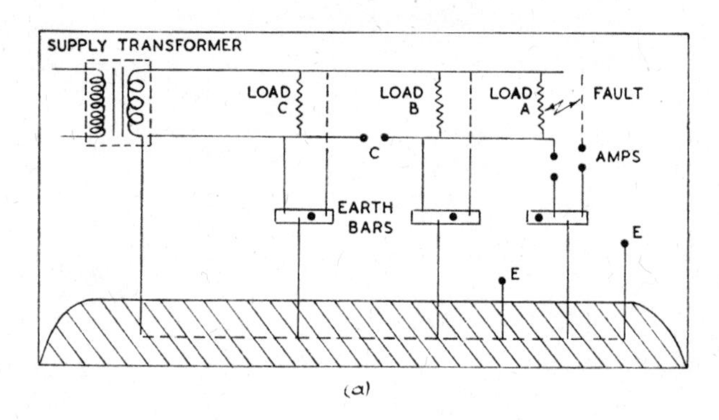

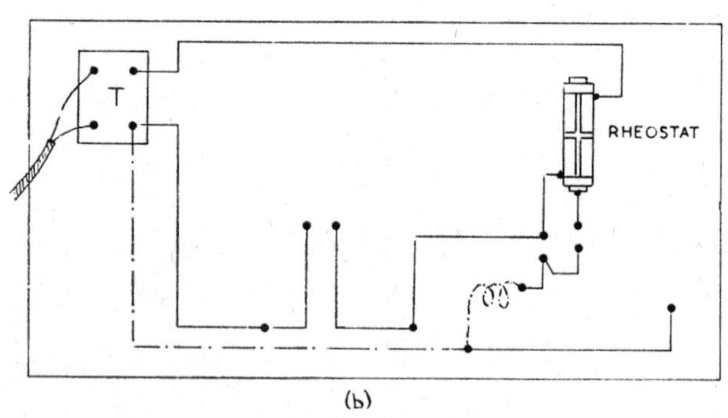

FIG. 14.4 (a–b) Demonstration Board No. 4
(a) Front view; (b) Rear view.

ondary winding and a 12-ohm rheostat are mounted on the rear of the board as shown in Fig. 14.4(b). As in demonstration board No. 3, the earth connections are each made using approximately 5 ft. of 20 S.W.G., D.C.C. nichrome resistance wire.

Demonstration Board No. 5. Load with Adjustable Earth Fault
The diagram is painted on the front of the board and terminals positioned as shown in Fig. 14.5. The load consists of a 250 ohms, 3 ampere rated enclosed type rheostat, the slider providing the 'earth fault'.

Demonstration Board No. 6. Polarity Test Board
The main switch, 5A one-way lighting switch, batten holder and earth bar are arranged as shown in Fig. 14.6. A 5A intermediate lighting switch, mounted on the rear of the board, serves to reverse the polarity of the lighting circuit as required.

Demonstration Board No. 7. Insulation Test Board
A 'consumer's unit' is mounted on the front of the board as shown in Fig. 14.7(a). The outgoing terminals of the consumer's unit are connected to terminal strips at the rear of the board as shown in

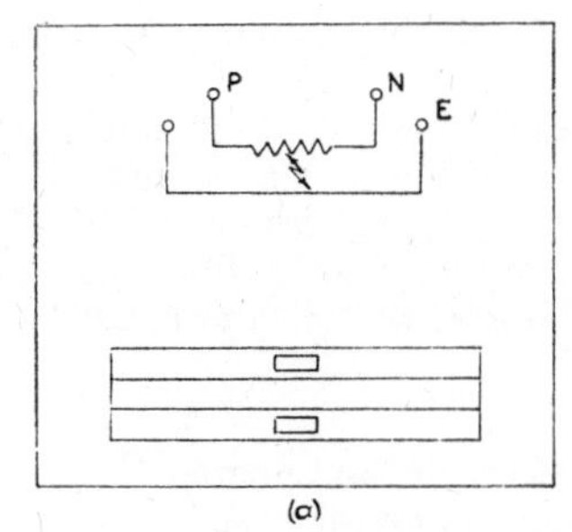

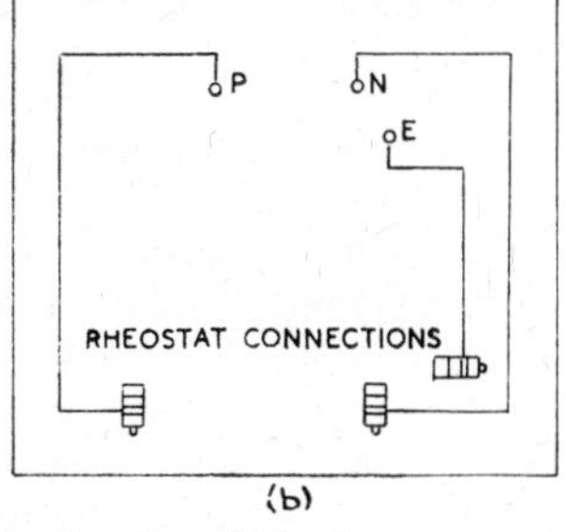

FIG. 14.5 (a–b) Demonstration Board No. 5
(a) Front view; (b) Rear view.

Fig. 14.7(b) so that radio type carbon resistors can be readily connected to simulate the insulation resistance between conductors and to earth as required. Values of resistance between 0·5MΩ and 10MΩ are very suitable for this purpose.

Demonstration Board No. 8. Effectiveness of Earth Test Board
A 13A socket outlet is mounted on the front of the board as shown in Fig. 14.8. A 5 ohm rheostat, capable of carrying 30A, is mounted on the rear of the board and connected in series with the earthing lead, so that the earth impedance may be varied at will. The intermediate switch, S, serves to reverse the polarity of the socket outlet when required.

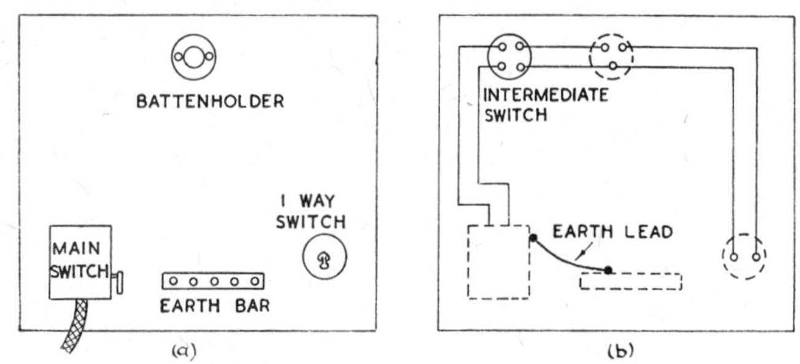

FIG. 14.6 (a–b) Demonstration Board No. 6
(a) Front view; (b) Rear View.

Demonstration Board No. 9. Earth Electrode Resistance Measurement
The diagram is painted on the front of the board, and 4 millimetre insulated sockets positioned as shown in Fig. 14.9(a). Connections are made as shown in Fig. 14.9(b) using approximately 3 ft. of 20 S.W.G. D.C.C. nichrome resistance wire for connections marked X, 1 ft. of the same wire for connections marked Y, and 1/·036 bell wire for all other connections.

Demonstration Board No. 10. Three-heat Switch Control
The three-heat switch and batten holders are arranged as shown in Fig. 14.10, the batten holder terminals being connected to substantial brass terminals as shown.

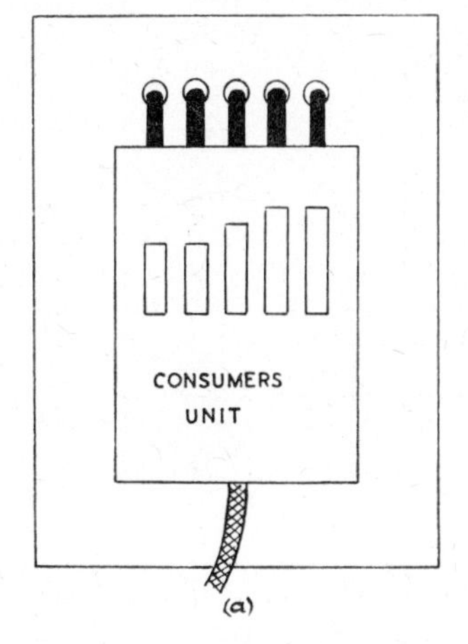

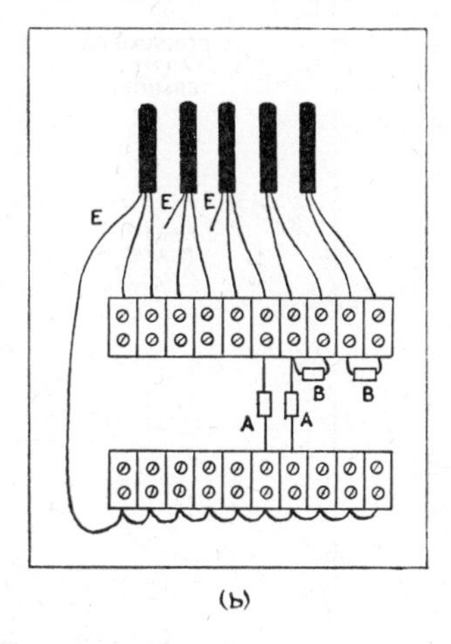

FIG. 14.7 (a–b) Demonstration Board No. 7
(a) Front View; (b) Rear View.

A = Resistors to 'Earth'.
B = Resistors between Phase and Neutral.
E = Earth Continuity Conductors (all connected to lower terminal strip).

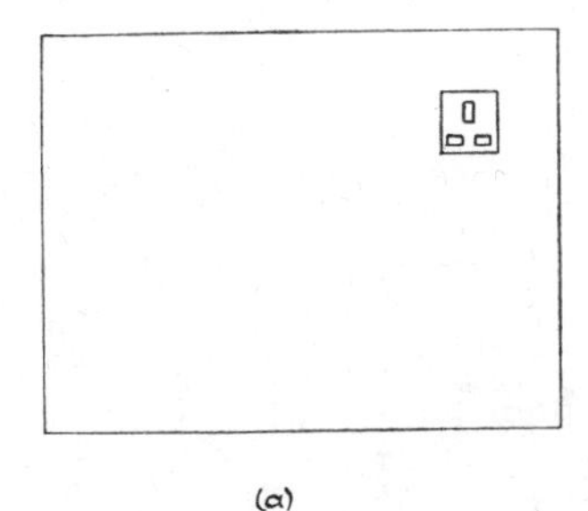

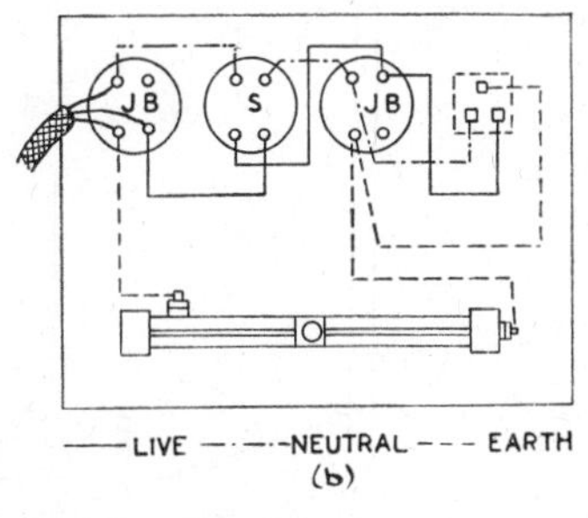

FIG. 14.8 (a–b) Demonstration Board No. 8
(a) Front view; (b) Rear view.
J.B. = Joint Box. S. = Intermediate Switch.

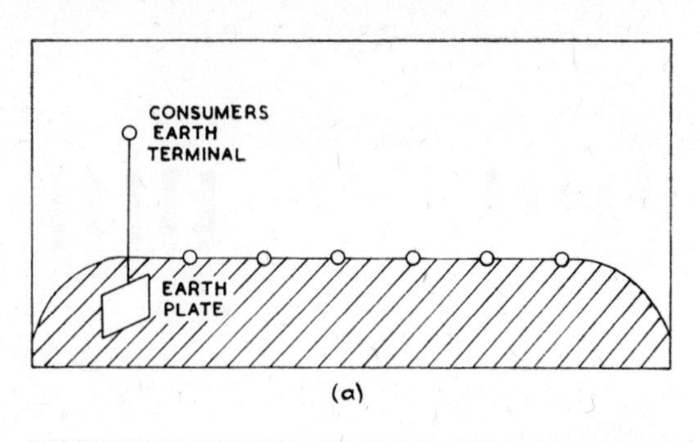

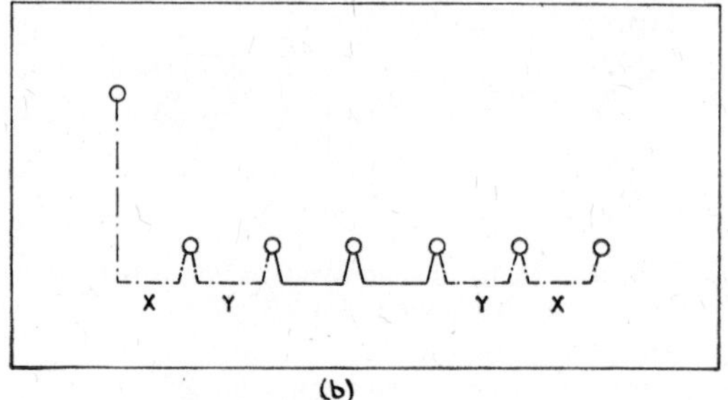

FIG. 14.9 (a–b) Demonstration Board No. 9
(a) Front view; (b) Rear view.

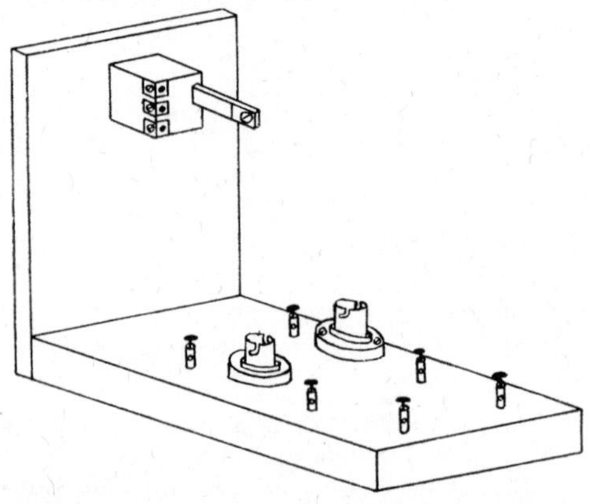

FIG. 14.10 Demonstration Board No. 10

Demonstration Board No. 11. Simmerstat Control

The simmerstat control switch and batten holder are arranged as shown in Fig. 14.11, the batten holder terminals being connected to substantial brass terminals as shown.

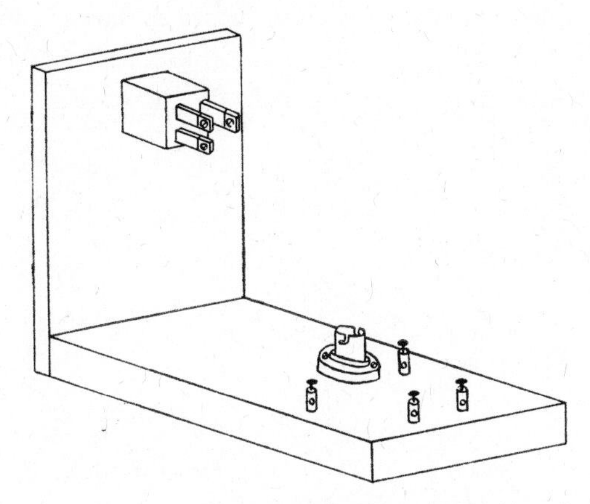

FIG. 14.11 Demonstration Board No. 11

Demonstration Board No. 12. Fluorescent Lamp (Switch Start)

The components are mounted on the board as shown in Fig. 14.12. The terminals of each individual component are connected to

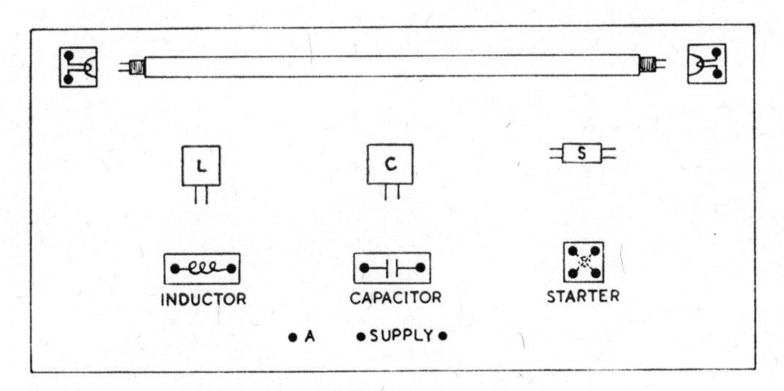

FIG. 14.12 Demonstration Board No. 12

substantial brass terminals, the theoretical symbol of the component being marked on each terminal block.

Demonstration Board No. 13. Fluorescent Lamp (Lead-Lag Circuit)

The components are mounted on the board as shown in Fig. 14.13.

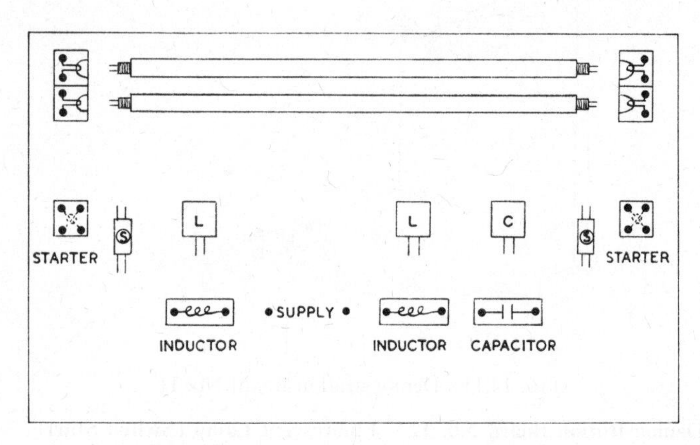

FIG. 14.13 Demonstration Board No. 13

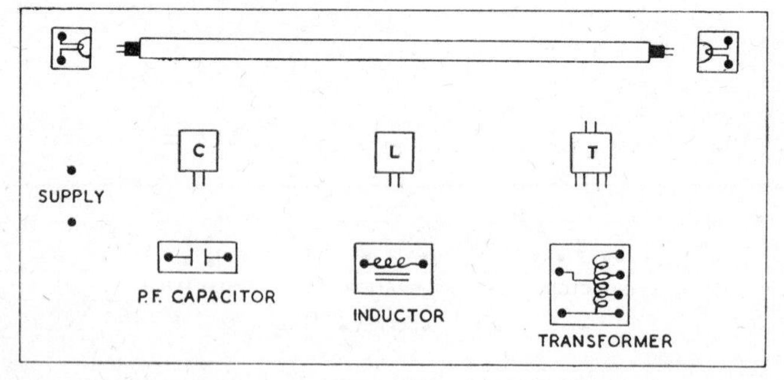

FIG. 14.14 Demonstration Board No. 14

The terminals of each individual component are connected to substantial brass terminals, the theoretical symbol of the component being marked on each terminal block.

Demonstration Board No. 14. Fluorescent Lamp (Instant Start)
The components are mounted on the board as shown in Fig. 14.14. The terminals of each individual component are connected to substantial brass terminals, the theoretical symbol of the component being marked on each terminal block.

Demonstration Board No. 15. High-pressure Mercury Vapour Lamp
The components are mounted on the board as shown in Fig. 14.15. The terminals of each individual component are connected to substantial brass terminals, the theoretical symbol of the component being marked on each terminal block.

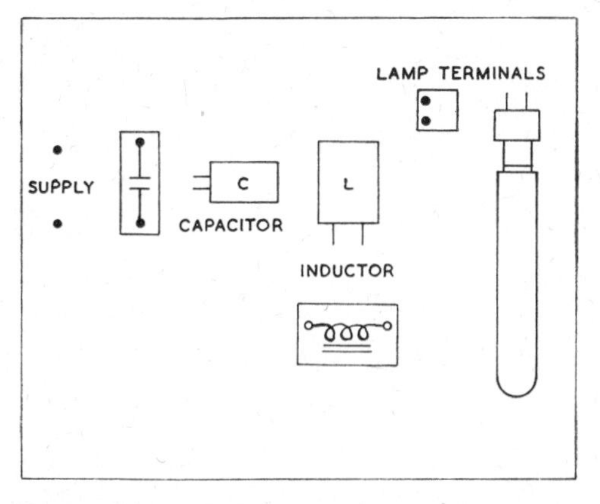

FIG. 14.15 Demonstration Board No. 15

Demonstration Board No. 16. Sodium Lamp
The components are mounted on the board as shown in Fig. 14.16. The terminals of each individual component are connected to substantial brass terminals, the theoretical symbol of the component being marked on each terminal block.

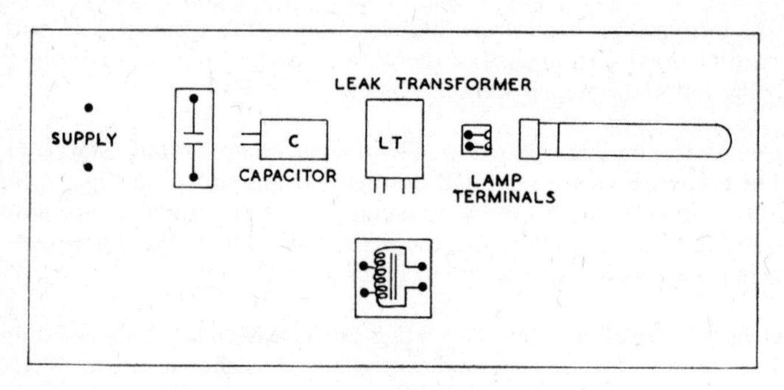

FIG. 14.16 Demonstration Board No. 16

CHAPTER 15

Typical Examination Questions

As far as possible the following questions have been grouped according to the chapter to which they mainly refer. However, in some cases the student may find it necessary to study more than one chapter in order to answer all parts of a particular question. Questions on wiring systems (Chapters 3 and 4) have been grouped together under one heading.

Chapter 1

1. (*a*) With the aid of a diagram explain the method of making up a lighting pendant consisting of twin-twisted braided flexible cord, two-plate ceiling rose and a lamp holder.

(*b*) Give the size of flexible cords which have the following current ratings: (i) 2A; (ii) 10A; (iii) 15A.

(U.E.I., 1963)

2. (*a*) Describe the method of preparing an insulated cable, sweating it into a lug or socket and finishing off. Illustrate your answer with sketches.

(*b*) What are the especial precautions quoted in regulations for this work?

(C.G.L.I., 1962)

3. With the aid of labelled sketches describe in detail how to make and insulate a married joint in 19/·064 cable.

(E.M.E.U., 1962)

4. Draw a pictorial sketch approximately full-size of a 13A plug-top, showing the position of the fuse and the colours of a three-core flexible cord connected to it.

State briefly the main points to observe when connecting a flexible cord to such a plug top.

(U.E.I., 1963)

5. Explain the meaning of the terms conductor and insulator as applied to the electrical properties of a material. Give three examples of each type and suggest an application for each material listed.

(U.L.C.I., 1963)

Chapter 2

1. (*a*) Define a ring circuit and give all requirements to satisfy I.E.E. Regulations for the Electrical Equipment of Buildings.

(*b*) Give the minimum cable ratings (both for stranded and mineral-insulated conductors) and the maximum fuse ratings for the following alternative arrangements of 13-ampere socket outlet installations:

(i) Two socket outlets as one sub-circuit.
(ii) Six socket outlets as one sub-circuit.
(iii) Ring circuit in industrial premises.
(iv) Ring circuit in flat or house.

(C.G.L.I., 1962)

2. Show, with the aid of a circuit diagram, how three lighting points looped together may be operated by two-way switch control. Indicate on your diagram the cable size and colours.

(N.C.T.E.C., 1963)

3. (*a*) What is the maximum voltage drop allowed by the I.E.E. Regulations in a lighting installation?

(*b*) Forty lighting points are to be installed in a furniture warehouse and wired with V.R.I. cables drawn into heavy-gauge screwed conduit. It can be assumed that each lighting point requires a 240V 150W lamp. Explain how you would decide the number of final sub-circuits which are required and also give the size of the cable and fuse link ratings.

(U.E.I., 1963)

4. (*a*) Explain the effect of ambient air temperature on the permissible current ratings of ordinary V.R.I. cables.

(*b*) State three factors affecting the permissible current ratings of cables.

(*c*) State the consideration that should be given to voltage drop in the assessment of cable sizes.

(*d*) What percentage effect will a voltage drop of 10% have on the output of a heater?

(C.G.L.I., 1955)

5. A hospital corridor is lit by two 100-watt lamps in parallel, controlled by two two-way switches. It is necessary to install a further point of control mid-way along the corridor using an intermediate switch.

Show by means of a circuit diagram:

(*a*) The original circuit.
(*b*) The circuit when modified.

(E.M.E.U., 1962)

6. A consumer requires a supply to a 2·5kW heater unit which is situated 30 yds. from the meter where the voltage is 250V. The I.E.E. Wiring Regulations limit the voltage drop in this type of installation to 1V plus 2% of the supply voltage. Determine whether or not 3/·036 cable will be suitable for this installation. The current rating of 3/·036 cable is 15A, and at this current a 1V drop occurs every 11 ft. of run.

(W.J.E.C., 1963)

7 (*a*) Draw a diagram showing the sequence of the control equipment to be installed at the supply intake to a house. Show which parts are the responsibility of the supply company and which are the responsibility of the consumer.

(*b*) Under what conditions may the consumer's Main Fuse be omitted?

(U.E.I., 1963)

Chapters 3 and 4

1. (*a*) Describe fully the precautions to be taken in constructing a metallic conduit installation.

(*b*) Describe the method of testing the resistance of such a conduit system, and state the acceptable values stipulated by the I.E.E. Regulations.

(C.G.L.I., 1960)

2. State in your own words the I.E.E. Regulations dealing with the corrosion of metal conduits and explain the precautions necessary to prevent corrosion of steel conduits.

(U.E.I., 1963)

3. (*a*) Why are sealing-boxes necessary when terminating paper-insulated cables?

(*b*) Sketch a box suitable for terminating a two-core paper-insulated lead-sheathed and armoured cable.

(C.G.L.I., 1961)

4. State the requirements for a temporary lighting installation in respect of the following:

(*a*) Anticipated use exceeding three months.
(*b*) The loading of final sub-circuits.
(*c*) Cable sheathing.
(*d*) Responsibility.
(*e*) Control.

(C.G.L.I., 1961)

5. State in your own words the requirements of the I.E.E. Regulations for the installation of T.R.S. cable concerning:

(*a*) Exposure to sunlight.
(*b*) Terminations at wiring outlets.
(*c*) Cable runs under floors.
(*d*) Proximity to gas pipes, etc.

(E.M.E.U., 1963)

6. State the respective advantages and disadvantages of the following systems of wiring:

(*a*) Screwed metal-conduit with P.V.C.-insulated cables.
(*b*) Tough rubber-sheathed cable.
(*c*) Lead-sheathed cable.

(U.E.I., 1963)

7. Describe, with the aid of sectional sketches of the cables, an earthed concentric system of wiring.

What are the advantages and disadvantages of this system compared with a normal conduit installation?

(U.E.I., 1963)

8. (*a*) What are the component parts of a mineral insulated metal-sheathed cable?

(*b*) For what particular conditions is this form of cable most suitable?

(C.G.L.I., 1961)

9. State the requirements of the I.E.E. Regulations covering the installation of metallic duct and trunking systems with particular reference to:

(i) Protection from mechanical damage.
(ii) Entries to the system.
(iii) Outlets from the system.
(iv) Joints and junctions.
(v) Bends.
(vi) Space factor.
(vii) Use on a.c. systems.

(C.G.L.I., 1960)

10. Describe in detail with the aid of a sketch, the construction of a four-core, paper-insulated, lead-covered, steel-wire-armoured cable.

(E.M.E.U., 1962)

Chapter 5

1. Describe a suitable form of protection for an installation against earth leakage suitable for use where an earth electrode of low impedance is not available.

Illustrate your answer with a wiring diagram of the equipment.

(C.G.L.I., 1960)

2. (*a*) Explain the precautions which should be taken when installing a voltage-operated earth leakage circuit-breaker.

(*b*) Draw the circuit diagram of the above, showing how it is connected in an installation.

(U.E.I., 1963)

3. A fuse is an essential part of an electric circuit. Explain:

(*a*) Its purpose.
(*b*) How it operates.
(*c*) Why it is important that the fuse element is of the correct rating.

(U.E.I., 1963)

4. (*a*) What is the smallest size conductor that may be used as an earthing lead?

(*b*) Describe a form of protection for an installation against earth leakage, suitable for use where an earth electrode of low impedance is not available. Illustrate the answer with a wiring diagram.

(W.J.E.C., 1963)

5. I.E.E. Regulation No. 406 covers the basic requirements for earthing. Complete in your own words the following which is an extract from this regulation: 'The earthing arrangements of the consumer's installation shall be such that a fault current (from a phase or non-earthed conductor to adjacent exposed metal) of . . .' 'Alternatively, and in every instance, where this requirement cannot be met . . .'

(E.M.E.U., 1962)

Chapter 6

1. Describe how you would carry out the following tests on an installation:

(*a*) Earth continuity test.
(*b*) Polarity test on single-pole switches.

(U.E.I., 1963)

2. Describe all the tests that should be applied to a completed installation. Give typical readings that would be obtained if the installation were in good order.

(N.C.T.E.C., 1963)

3. What is meant by the term 'insulation resistance' of an electrical installation? State the regulations pertaining to insulation resistance to earth of domestic installations. Describe how a completed installation is tested to ascertain if it complies with I.E.E. Regulations regarding insulation resistance to earth.

(W.J.E.C., 1963)

4. A conduit run 100 ft. long contains six single-core cables, one of which is open-circuited.

Describe how to locate this fault using a continuity tester.

(E.M.E.U., 1962)

5. (*a*) Describe the procedure for testing the insulation resistance of an installation.

(*b*) What are the minimum values prescribed in I.E.E. Regulations?

(C.G.L.I., 1963)

6. Describe the method of earth loop testing. Illustrate your answer with a diagram of a method employing current injection equipment.

(C.G.L.I., 1958)

Chapter 7

1. (*a*) What do you understand the following terms to mean:

(i) Pressure-type water-heater.
(ii) Free-outlet water-heater.

(*b*) Explain in each case how the heated water is delivered to the outlet point, illustrating your answer with sketches.

(C.G.L.I., 1962)

2. (*a*) Show, by separate drawings for 'High', 'Medium' and 'Low' heat positions, the connections of a series-parallel switch controlling two separate sections of resistance wire forming the element of a heating appliance.

(*b*) Assuming the two sections of resistance wire are of equal resistance, what is the proportional current flow and heating effect in the 'Medium' and 'Low' positions relative to the 'High' position.

(C.G.L.I., 1962)

3. Describe, with the aid of sketches, the construction and principle of operation of:

(*a*) A radiant heater.
(*b*) A convector heater.

List the advantages of each type.

(W.J.E.C., 1963)

4. Describe, with sketches, two different types of electric room heaters, explaining clearly the working principles of each.

(C.G.L.I., 1959)

5. Sketch and describe the type of heater you would install in the hall of a house. Give reasons for your choice.

(U.E.I., 1963)

6. (*a*) By means of a suitable labelled sketch, show the construction of a thermostat which is suitable for an immersion heater.

(*b*) Draw the circuit diagram for an immersion heater including the thermostat connections which is supplied from a double-pole switched fused spur-box fitted with a 13A fuse.

(U.E.I., 1963)

Chapter 8

1. (*a*) Draw a circuit diagram of a hot-cathode fluorescent lamp with inductive ballast and starter switch control.

(*b*) Describe the operation of the circuit.

(*c*) What is meant by stroboscopic effect and how can this be minimized?

(C.G.L.I., 1958)

2. What do the I.E.E. Regulations require with regard to fluorescent lighting installations for the following?

(i) Protection of live parts.
(ii) Switches.
(iii) Loading of sub-circuits.

(C.G.L.I., 1959)

3. State the requirements of the I.E.E. Regulations concerning the installation of fluorescent discharge lamps in reference to:

(*a*) Control switches.
(*b*) Loading of final sub-circuits.
(*c*) Mounting of ancillary apparatus.

Draw a diagram of connections for a final sub-circuit comprising one fluorescent lamp, starting-switch, inductor, capacitor and control switch.

(C.G.L.I., 1951)

Chapter 9

1. Draw the circuit diagrams for the following:

(*a*) One trembler bell rung from any one of four pushes through a four-way indicator board having pendulum type elements. Show the internal connections of the bell.

(*b*) One lighting point controlled from any of three positions. Indicate the colour of the conductors.

(U.E.I., 1963)

2. (*a*) What do you understand is meant by the phrase 'closed-circuit' working and what are the advantages when used in alarm and signal installations?

(*b*) Draw a diagram of a simple closed-circuit burglar alarm system covering four windows, two doors and one safe. The alarm to be given by a bell and the system to be provided with a 'day' cut-out.

(C.G.L.I., 1960)

3. (*a*) Neatly sketch an electric bell of the trembler type, clearly showing all components.

(*b*) Describe the action of the bell in producing repeated hammer strokes.

(C.G.L.I., 1960)

4. Explain the function of a relay. Illustrate your answer by drawing a circuit which contains a relay.

(U.E.I., 1963)

5. Describe, with the aid of a free-hand sketch, the construction of a telephone transmitter (microphone). Describe the action of the microphone in converting the varying sound wave pressures into a varying electric current.

(E.M.E.U., 1962)

6. An extra low voltage bell is to be operated from four positions through a four-way indicator board. The supply is to be taken from a 240V a.c. supply via a bell transformer. Draw a circuit diagram of the whole installation and the necessary control equipment. Describe, with reference to the diagram, how this installation can comply with the I.E.E. Regulations.

(E.M.E.U., 1962)

7. Draw a connection diagram for a simple two-instrument telephone set working between two offices. The instruments are to allow ringing and speaking from both ends.

Explain the operation of the circuit.

(C.G.L.I., 1962)

8. An alarm bell is situated a long distance from the bell-push, and a relay is necessary.

Draw a connection diagram of a bell and battery circuit, including a relay, and explain the working of the circuit.

Sketch and describe the relay.

(C.G.L.I., 1962)

9. Describe, with the aid of clear sketches, the operation of the following parts of an ordinary telephone:

(*a*) The carbon microphone.
(*b*) The receiver.

(C.G.L.I., 1962)

Chapter 10

1. Discuss briefly the two different methods of charging secondary batteries, stating the particular advantages of each method.

(N.C.T.E.C., 1963)

2. What tests and observations can be made on a battery of lead-acid cells to determine their condition and state of charge? Describe the conditions necessary in a room in which a battery of lead-acid cells are housed.

(W.J.E.C., 1963)

3. (*a*) Briefly compare the properties and applications of lead-acid and alkaline secondary cells.

(*b*) Sketch the graph showing how the voltage of a lead-acid cell varies during charge and discharge.

(*c*) Calculate the charging current when a battery of e.m.f. 11·7V and negligible internal resistance is connected to a 16V supply with a series resistor of 1·5 ohm.

(E.M.E.U., 1963)

4. (*a*) What tests and observations would you make on a battery of lead-acid cells to determine their condition and state of charge?

(*b*) Describe the conditions necessary in a room or building in which a battery of lead-acid cells are housed.

(C.G.L.I., 1961)

5. Describe with diagrams EITHER the constant-voltage OR the constant-current methods of charging secondary batteries.

Forty lead-acid secondary cells are charged at a constant rate of 12 amperes for ten hours. The supply is 240V direct current. The e.m.f. of each cell at the beginning of the charge is 2·0V, and at the end of charge the e.m.f. is 2·7V.

Calculate the values of the variable charging resistance at beginning and end of charge. Ignore the internal resistances of the cells. How would you check that the cells were fully charged?

(C.G.L.I., 1956)

Chapter 11

1. Make a drawing showing the connections of a shunt-wound d.c. motor and starter, complete with the usual protective devices, and explain how you would reverse the rotation of the motor.

(C.G.L.I., 1957)

2. Explain, with diagrams, how shunt, series and compound wound generators differ from each other. Sketch graphs showing the variations of terminal voltage with load current for each type of generator giving reasons for the shapes of the graphs. State briefly a use for each type of generator in practice.

(U.L.C.I., 1963)

3. What is the meaning of the expression back e.m.f. of a direct current motor?

Explain how the back e.m.f. and the current change during the starting of a d.c. motor.

(C.G.L.I., 1962)

4. What do you understand by the term 'back e.m.f.' of a d.c. motor? Explain how back e.m.f. affects the starting of a d.c. motor.

A d.c. motor connected to a 460V supply has armature resistance of 0·15 ohm. Calculate:

(*a*) The value of the back e.m.f. when the armature current is 120A.
(*b*) The value of the armature current when the back e.m.f. is 447·4V.

(C.G.L.I., 1956)

Chapter 12

1. Draw the circuit of a direct-on-line contactor starter suitable for a 2-h.p. three-phase induction motor. An emergency 'stop' button fitted at a remote position must be included in the circuit.

(U.E.I., 1963)

2. (*a*) Discuss the advantages and disadvantages of Cage Rotor Motors for general industrial use.

(*b*) Give two examples of conditions where special enclosures are required to protect the motor and the type of enclosure you would select.

(*c*) Deduce a formula for determining the synchronous speed in r.p.m. of an induction motor in terms of the number of poles and the frequency of the supply.

(C.G.L.I., 1962)

3. What do you understand by the slip of an induction motor?

Calculate the speed in rev/min of a six-pole induction motor which has a slip of 6% at full load, with a supply frequency of 50 c/s.

What will be the speed of a four-pole alternator supplying the motor?

(C.G.L.I., 1961)

4. (*a*) Describe the rotor-resistance method of starting a three-phase slip-ring induction motor. Give a diagram of connections to include the motor windings.

(*b*) Find the speed in rev/min of a four-pole induction motor which is running with a slip of 6%. The frequency of the supply to the motor is 50 c/s.

(C.G.L.I., 1963)

Chapter 13

1. (*a*) Describe one method of fixing a motor where adjustment is required for correcting belt tension.

(*b*) Describe a suitable means for making final connections to the motor in these circumstances and illustrate your answer with a sketch.

(*c*) What control apparatus is specified in the I.E.E. Regulations for motors exceeding $\frac{1}{2}$ horse-power?

(C.G.L.I., 1961)

2. Describe, with the aid of a diagram, how to test the armature of a d.c. generator for faults which occur in service.

Explain how these faults show up in your tests.

(E.M.E.U., 1962)

3. Give the requirements of the I.E.E. Regulations covering the installation of motors in relation to:

(i) Voltage drop on motor supply cables.
(ii) Fusing.
(iii) No-volt protection.
(iv) Overload protection.

(C.G.L.I., 1960)

4. A small portable air compressor is driven by a petrol engine through a vee-belt drive. Describe a method of aligning and fixing an electric motor which is, to replace the petrol engine.

(U.E.I., 1963)

Solutions to Numerical Examples

Chapter 2

4(*d*) 19%.

6. Actual current = 10A
   Actual voltage drop = 6V
   Permitted voltage drop = 5·46V

Chapter 10

3(*c*). 2·87A.

5. 13·3Ω, 11Ω.

Chapter 11

4. (*a*) 442V (*b*) 84A

Chapter 12

3. 940 r.p.m. 1,500 r.p.m.
4. 1,410 r.p.m.

Index

Index

S

T

V

W